Network Layer
Switched Services

Network Layer Switched Services

Daniel Minoli
Andrew Schmidt

WILEY COMPUTER PUBLISHING

JOHN WILEY & SONS, INC.

New York • Chichester • Weinheim • Brisbane • Singapore • Toronto

Publisher: Robert Ipsen
Editor: Marjorie Spencer
Assistant Editor: Kathryn Malm
Managing Editor: Erin Singletary
Text Design & Composition: North Market Street Graphics

Designations used by companies to distinguish their products are often claimed as trademarks. In all instances where John Wiley & Sons, Inc., is aware of a claim, the product names appear in initial cap or ALL CAPITAL LETTERS. Readers, however, should contact the appropriate companies for more complete information regarding trademarks and registration.

This book is printed on acid-free paper. ∞

Published by John Wiley and Sons, Inc.

Published simultaneously in Canada.

Library of Congress Cataloging-in-Publication Data:
Minoli, Daniel, 1952–
        Network Layer Switched Services / Daniel Minoli, Andrew Schmidt.
            p.   cm.
        ISBN 0-471-19080-2 (alk. paper)
        1. Asynchronous transfer mode.   2. Internetworking
(Telecommunication)   3. Computer network protocols.   I. Schmidt,
Andrew.   II. Title.
TK5105.35.M38   1998
004.6'6—dc21   97-31834
                                                                          CIP

Printed in the United States of America.

10 9 8 7 6 5 4 3 2 1

Contents

Preface

What motivates the need to migrate? In telegraphic terms, first there was the mainframe. Then there was the desktop. Then there was the enterprise network. Next came the commercialized Internet and corporate intranets. This synthesis has evolved to the point where, as some say, "the computer is the network." But, in fact, we are now entering a quantum-change phase in the corporate landscape, whereby *"the corporation is the network."*

Not in any way perturbed by the furious rate at which change is occurring, intrepid communication planners make every effort to embrace the evolving network technologies, which give the promise to offer organizations ever-increasing options, productivity gains, and support for a total transition to information-based economies and electronic commerce.

This book aims to provide at an early time a thorough description of the technologies that will see deployment in progressive organizations in the next couple of years. Eventually, all networks could look like those described in this text.

So, we hope to spark an intellectual scintilla and, if we are so lucky, give life to some enlightenment.

Dan Minoli
Andrew Schmidt

Acknowledgments

The authors would like to thank Fore for the input received, particularly in Chapters 3 and 4. The authors would like to thank 3Com for the input received in Chapter 2. Mr. C. Semeria, Mr. D. Flyn, and Ms. D. Christensen of 3Com are thanked for material and insight provided. The authors also would like to thank the Fibre Channel Association for the material used in Chapter 8.

Mr. Minoli also thanks Mr. Ben Occhiogrosso, President, DVI Communications, for his suggestions and assistance.

Network Layer
Switched Services

1 Overview of Evolving Switching Technologies

This book is about the evolving networking technologies that will be important for turn-of-the-century networks, with emphasis on generalized switching technology.

Introduction and Background

Switching is the fundamental ability to connect two or more parties without having to dedicate a path between them on a permanent basis. Our definition of switching is broad enough to include functions that are not always considered to be pure switching in current parlance. Nonetheless, in minimalistic terms such functions are really switching functions, and it is worth looking at these as being switching functions. Figures 1.1 and 1.2 depict the concept of switching at the connectivity level—Figure 1.1 shows what would be required without switching, while Figure 1.2 shows the advantages and components of a switched network.

Figure 1.3 depicts important technologies now available; many of these are covered in this book. As can be seen in the figure, some technologies are applicable to the public network, others are applicable to both public and private networks, and yet others are mostly applicable to private networks.

Switching can occur at every layer of the Open System Interconnection Reference Model (OSIRM) for protocol architectures.

Switching can occur at the physical layer. For example, dial-up connectivity is an example of circuit switching. Not every telephone (or modem)

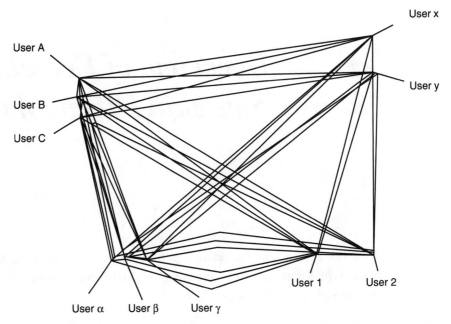

Figure 1.1 *Without switching, every user is connected directly to every other user (here, $10 \times 9/2 = 45$ links).*

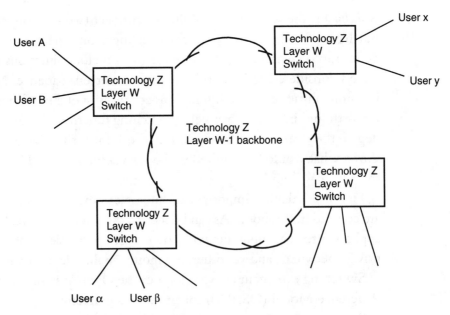

Figure 1.2 *Switching simplifies connectivity between users.*

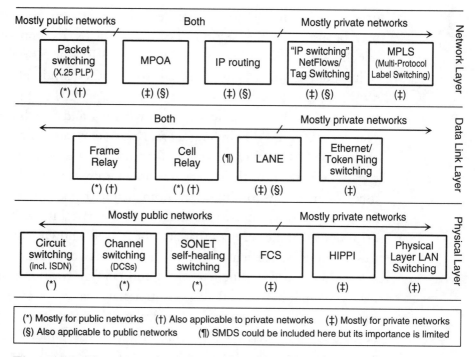

Figure 1.3 *Taxonomy of switching technologies.*

user is connected to every possible destination. Instead, every user is equipped with a physical link (actually residing below the physical layer) to a local access point (e.g., a telephone company or mobile telephone company central office). The destinations are connected to remote access points. A backbone network, or collection of networks, exists that connects, in some appropriate fashion, all the access points—but not, a priori, the end users. When the local user, which is physically connected in a permanent fashion to the access point, desires to connect to a remote destination, it signals the local switch for service and, upon supplying the address of the remote destination, it is connected for the duration of the session to the intended destination. To achieve this, the backbone network supplies an appropriate backbone facility out of its pool of available communication resources to the origination-destination pair for the duration of the session. This resource is adjoined to the permanent link that the origination and destination users have with their respective access points to achieve an end-to-end path. For the next session, the local user may need to be connected to another destination; different backbone facilities may be used for this session. The switching is accomplished by such telephony switches as Lucent Technologies' 5 ESS or Nortel's DMS 500.

In addition to dial-up, there are some other kinds of physical layer switching that may be of interest to the end user. For public networks, these are digital crossconnect system switching and Synchronous Optical Network (SONET)[1] switching. *Digital crossconnects* are slower switches that enable carriers to alter the physical paths that are used in the access or backbone portion of their networks on a semipermanent basis. These are used for manually routing away from major trunk failures or for other kinds of quasi-static network redesigns. In the past this was known as *channel switching*. Some services that are not as well-known have come out of this technology, particularly in the area of network management and in restoration of private line (e.g., DS1 and DS3) networks. *SONET switching* is accomplished by add/drop multiplexors that can provide self-healing capabilities. These systems automatically route (or switch) away from a failed fiberoptic link, on the assumption that the underlying topology is physical ring–based, which is now the case in many carriers' networks.

Circuit switching, channel switching, and SONET switching will not be treated further in this book, since there is ample literature on these topics. *Fibre Channel* (FC) technology, shown in Figure 1.3, is treated due to its relative novelty and also because it has both processor-level and local area network (LAN) applications.[2] Evolving gigabit LANs will be based on FC technology. FC capabilities include switching fabrics that support gigabit-level connectivity among users on the network.

Switching can occur at the data link layer of the OSIRM. Well-known examples include frame relay (FR), Asynchronous Transfer Mode (ATM), Ethernet switching, and LAN emulation (LANE). *Frame relay* allows data users to be interconnected without requiring pairwise facilities between each user, just as described in the previous paragraphs. The interconnection speed is up to 1.536 Mbps. *ATM* does the same, but the interconnection speed is currently up to 622 Mbps; in addition, ATM supports a

---

[1] Outside the United States, the digital hierarchy in question is called *Synchronous Digital Hierarchy* (SDH).

[2] Because FC has a framing layer, one could also classify it as a data link layer protocol. For this discussion it is treated as a physical layer protocol.

variety of quality of service (QoS) options, making it an ideal medium to support multimedia applications. These technologies apply principally to public networks but can also be found in private network applications.

Ethernet switching allows Ethernet LAN users to be interconnected in an effective manner. Initially, Ethernet employed a form of implicit or distributed switching. Users were connected over a shared common cable. Information meant for another user was labeled with the address of the intended recipient and injected onto the cable. Every user in the LAN received the data, but only the intended user accepted it and processed it further. This achieved switching (the ability to connect every user pair) in a distributed fashion. However, because of speed limitations of the interconnecting cable—especially as the number of users increased, so that the bandwidth per user decreased—segmentation was utilized. With segmentation, the entire community of users is partitioned into smaller sets. Each set of users now utilizes a LAN segment, thereby increasing the available bandwidth per user. In turn, LAN segments are interconnected—using, in effect, a multiport bridge. In order to support multimedia, the industry has gone in the direction of placing a single LAN user per segment. The segments are then interconnected by a high-end multiport bridge[3] called a *LAN switch*. The LAN switch achieves intracommunity connectivity without requiring each user to be connected with all others. These LAN switches are also called *Layer 2 switches* (L2Ss), because they operate at the medium access control (MAC) level, which reaches into the data link layer (Layer 2) of the OSIRM. As an elaboration of this concept, given subsets of these one-user segments want to be recognized as a virtual community. The LAN switch can be used to provide this virtual community association, thereby supporting a *virtual LAN* (VLAN) concept.

Some corporate users may have deployed ATM technology to their desktops, in order to obtain higher capacity. Still, there is the need to interconnect legacy users in a seamless manner, so that they can freely communicate and share information. *LANE* is a technology that allows this interconnection (hence, it is classified here as a switching technology). LANE is typically implemented in L2Ss, because it is a bridging technol-

---

[3] A bridge can be viewed as a data link layer switch that interconnects a community (segment *A*) with another community (segment *B*); in so doing, it enables user *x* in community *A* to interconnect with user *y* in community *B*.

ogy. To support the interconnection the L2S must implement some software to support connection oriented–to–connectionless mode conversion as well as a LAN-to-ATM address conversion. An L2S implementing this software is called a *LAN emulation client* (LEC).

Switching can occur at the network layer of the OSIRM. Packet switching (based on the X.25 interface specification) is an example. A router—say, supporting the Internet Protocol (IP)—can be viewed as providing a switching function, although the jargon used to describe this function is *routing;* see Figure 1.4. In fact, in the past few years, routers have utilized a two-module internal architecture, where one module handles the route-determination function and the other module supports the forwarding (switching) function.

Until recently, routers were required to use a backbone of dedicated private lines, in order to achieve downstream supported user–to–downstream supported user connectivity. This can become impractical when the number of sites to be connected becomes large. Hence, the possibility of using a data link layer switching technology has been investigated. Frame relay connections can be used, as can ATM, in place of a backbone of dedicated lines. In effect, the virtual private lines provided by frame relay or ATM become the virtual backbone of this evolved network. Because ATM provides more bandwidth and has better QoS support, the industry emphasis now is on ATM. In this application, ATM can be used as a *fat pipe,* where it is seen strictly as a data link layer channel (which, in

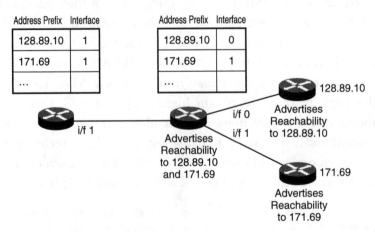

Figure 1.4 Destination-based routing.

effect, appears to the IP driver as a physical link). But to make better use of the ATM features, one needs more direct access to the ATM service capabilities. *Multiprotocol over ATM* (MPOA) is a recent approach whereby the router touches the ATM cloud in a more direct way, while it simultaneously achieves a one-hop distance from other routers at the edge—hence, ATM becomes a one-hop backbone. While some networks are being deployed today with LANE, the ATM Forum's MPOA work builds an architecture to fully integrate bridging (i.e., LANE), routing, VLANs, and ATM. MPOA also supports protocols other than IP.

The newly evolving technology known as *IP switching* crystallizes the switching function even better than is done on modern routers. We have already indicated that routers have segregated the switching function to a dedicated processor. However, that processor does not use ATM internally, and to use ATM externally it must still go through a stage of protocol data unit (PDU) conversion. IP switching goes all the way to using a distinguishable ATM switch to support the switching function. Long flows receive special treatment. *Tag switching* is an elaboration of IP switching by some established vendors.

Figure 1.2 implies that the backbone typically supports facilities one layer below the switching technology. Naturally, the user of switching can be recursive, as is shown in Figure 1.5. For example, an IP router supporting an IP-based network may use ATM to provide connectivity. In turn, ATM can use SONET links that use add/drop multiplexers to switch around failed fiber links.

"The Corporation Is the Network" Paradigm

It is clear that we have reached a stage where *the corporation is the network.*

Businesses have been looking for ways to increase their market shares and profits since antiquity. Over the centuries, even the millennia, businesses have successfully sought to use advances in technology to introduce and promote their products. The development of money was one of the key milestones along this continuum. This advance meant that instead of having to carry around bushels of grain for barter, our ancestors were able to exchange a few small disks of metal for their material and physical

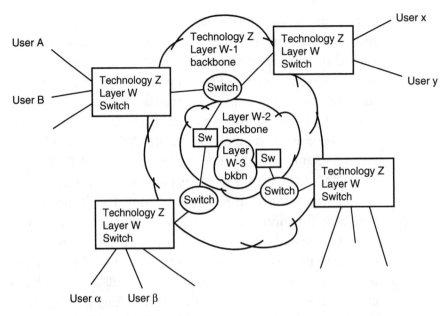

Figure 1.5 *Hierarchical switching.*

necessities. More recently, paper money came into use, as the use of large amounts of coins became inconvenient. During the past century, the telephone, fax, and electronic mail have provided faster, cheaper, and more reliable communication of business data within and between commercial entities. Geographical distances and multiple time zones are no longer barriers to business communications. The search for more efficient ways of doing business is now driving another revolution in the use of currencies and our concept of money. This revolution is known as *electronic commerce.*

Since the Industrial Revolution, many if not most businesses have been built in the proximity of what can be called *channels of goods distribution.* Initially, cities and businesses were built close by harbors or rivers. Later they were built close to railroads. Later yet, there were built within easy reach of the highway system. Just in the past decade or so, some goods-intensive businesses have placed themselves close to airports. Now what has become critical is access to a worldwide communication network. Electronic commerce can be defined as any purchasing or selling through a telecommunications network. At the same time, the nature of what is a manufactured good is changing—such a good either is simply information (e.g., a cash transfer; a credit given; a reservation established; an analysis, newsletter, or lecture; an electronic order received; or a downloaded movie or musical clip) or is highly dependent on information (e.g., a just-in-time manufacturing order or inventory transaction). Already, for a large number of companies, the product is an electronic artifact, what can be called an *e-product*. Unlike the old money made of stones, metal, and paper, the new money is kept in vaults of magnetic particles, and commerce is undertaken electronically.

Electronic commerce in support of e-products, or even real products, is becoming an established way of doing business. Electronic commerce is the symbiotic integration of communications, data management, and security services to allow business applications within different organizations to automatically exchange information.

The 1990s have seen the rise of a new form of industrial organization—the *networked firm,* sometimes known as the *virtual organization.* We synthesize this business predicament with the phrases "The corporation has become the network," or "The corporation is the network." Rather than being based on the rigid hierarchical internalized form of the traditional organization, the virtual company is based on a network paradigm. Use of networks enables the virtual firm to focus on core competencies, to engage in outsourcing or cosourcing of noncritical administrative functions, and to make large-scale use of a constantly shifting pool of subcontractors.

Therefore, the corporation is as good as, can be seen as, and can be identified with *its network.* Its outreach to the world and the inreach to it from any customer, anywhere, anytime, are totally defined by its network. Today, the network is the channel of goods distribution. Just as having access to distribution channels was important in the past, having a global-reach network is becoming a critical business survival issue today. Naturally, that network does not have to be wholly owned by the corporation, in much the same way that a road, an airport, or a railroad is not usually owned by the corporation. That network may be comprised of: (1) a *traditional enterprise network*—the physical foundation of the corporation's intracompany communication facilities; (2) the *intranet*—an overlay to the enterprise network that is a way to build uniform applications, clients, and servers that have the look and feel of Internet applications; (3) the *Internet;* (4) other *intercompany specialized networks*—for example, the NYCE banking network; and (5) *international extensions.* It is desirable that this fundamental, company-distinguishing synthesis of communication facilities—what can be called an omninet—also carries voice, video, image, and other media, in addition to the traditional data objects.

The Scope of This Book

Planners are now faced with the challenge, even the opportunity, of assessing which of the many technologies that have appeared in the 1990s will be ideal for turn-of-the-century networks in their own corporate environments. Planners are expected to make every effort to embrace the evolving network technologies, which promise to offer organizations ever-increasing connectivity options, productivity gains, and support for a total transition to information-based economies and electronic commerce.

This book aims to provide, at an early date, a thorough description of the technologies that will see deployment in progressive organizations in the next several years. Eventually, all networks could look like those described in this text. The book covers Ethernet switching (at 10 Mbps or higher, including gigabit rates), ATM, LANE, MPOA, IP switching, and FC technology.

Ethernet Switching

LAN switching has become an important technology that extends the useful life of Ethernet technology, particularly for nonvideo applications, although video can also be supported, to a limited extent, if needed. LAN switching supports microsegmentation of LANs to network segments supporting one user per segment, and then enables the interconnection of these subnetwork segments into one network supporting any-to-any communication.

New applications have often overtaxed LANs. This causes segments to be utilized at rates up to 50 percent, with bursts up to 80 percent, which is undesirable—particularly in Ethernet environments, where saturation is theoretically possible. (In practice, many keep their LAN utilization in the 30 to 40 percent range, and even lower in some cases). The use of LAN switches, in conjunction with VLANs, can reduce this utilization to less than 5 percent. This approach maximizes LAN availability and minimizes delay. Collapsed backbone routers do reduce design and management challenges, but the problems are that this approach does not scale effectively and that subnetworks are statically interconnected at the physical level. Scaling restrictions arise from the fact that once the port capacity is reached or the backplane bandwidth is exhausted, more routers are needed; the issue then is how to connect these routers among themselves. Invariably, a connection operating at a slower speed than the router's bus is, by necessity, required.

Both to secure relief from Moves/Adds/Changes activities and to secure increased bandwidth, planners should consider replacing or upgrading shared hubs with Ethernet switches, particularly for the more bandwidth-intensive and project-mobile users. The manager can then connect the switch-based users to less volatile network segments in the newly configured shared hub. Switching hubs support several hundred Mbps and several dozen Ethernets. They provide 10 to 100 Mbps of dedicated bandwidth to users. This approach increases server throughput by up to 50 percent and decreases network contention by up to 30 percent. Typically, switching hubs support high cross-segment throughput. VLANs or LANE/MPOA are the next step in the evolution, by providing higher performance, higher throughput, and better realization of the virtual connectivity capabilities described earlier.

Router and hub vendors have merged and partnered in the mid-1990s to facilitate the delivery of switched LAN products. Positive consequences of this industry consolidation are lower cost per port and better network management support across an entire technology product line (e.g., across the hub, the switch, the router, and the firewall). LAN switching can both coexist with ATM and be a migration step to ATM: The ATM backbone can connect Ethernet switches that support users and servers; at a later time, clients and servers can migrate to ATM. High penetration rates are predicted for switched Ethernet for the late 1990s.

Token Ring switching technology also appeared in the mid-1990s, although it lagged behind the emergence of switched Ethernet by a couple of years. It provides the same kinds of benefits to Token Ring environments that switched Ethernet provides: Token Ring switching utilizes

Layer 2 switching to increase the available desktop and server-level bandwidth. This switching technology can increase the LAN bandwidth per connection without requiring new workstation adapters, new wiring, or changes to existing workgroups or backbones.

100-Mbps Ethernet products are now being deployed in organizations (this technology is also known as *fast Ethernet*). After a number of proposals emerged in the early 1990s, the industry settled on two standards: 100Base-T and 100VG-AnyLAN. *100Base-T* uses the same Carrier Sense Multiple Access/Collision Detect (CSMA/CD) technique as traditional Ethernet; *100VG-AnyLAN* uses a more sophisticated, but incompatible, bandwidth-on-demand technique. The technology is simple and, for 100Base-T, requires no changes to applications or to the premises wiring if Category 5 wiring is already in place. The cost per port is attractive ($200 to $400, much less than for ATM and FDDI), and there is no retraining required for network management. On the negative side, however, the technology entails shared bandwidth, which is not ideal for multimedia, digital video, or LAN-resident voice. For example, basic-quality digital video applications require 1.5 to 6 Mbps per user. Since the utilization of a CSMA/CD system needs to be around 30 percent, at the practical level this implies that only about 10 to 20 users can be placed onto a 100Base-T segment before *switched* 100-Mbps systems have to be employed. Another limitation is that 100-Mbps Ethernet is not scalable; in addition, it is a broadcast technology and is, therefore, not as secure.

100VG-AnyLAN equipment is available at press time, although the market potential is anything but certain, since it needs new adapters, hubs, and four-pair Category 3 wiring. The hope of vendors is that if they can price the 10/100 adapters in the same range as 10-Mbps adapters, user organizations will deploy the technology. At press time, prices are $300 for autosensing 10/100-Mbps adapters, $250 per repeater port, and $3750 for 15-port hubs. A handful of vendors also manufacture 10/100-Mbps bridges, allowing workgroups to operate at 100 Mbps locally and to be interconnected to the rest of the organization at 10 Mbps (note that this is a reversal of the classical design method of making the backbone 100 Mbps and the ancillary networks 10 Mbps). The Fast Ethernet Alliance is taking the position that products will soon be in abundance. They predict that, given the low cost of fast Ethernet silicon chips, several dozen vendors may have products by press time.

As noted, switched 100-Mbps systems are also being introduced. This technology combines the higher-speed LAN technology (e.g., 100Base-T) and switching hubs with high backplane capacity. Switching limits the number of users per port to as few as one, thereby reducing the contention domain.

This topic is discussed further in Chapter 2.

ATM and ATM Switching

ATM is an enabling technology, delivering capacity that could change the nature of networking at both the local and the wide area levels. ATM[4] technology takes full advantage of the large capacity potential fiber optic media. Interworking with legacy networks and protocols is a feature that has been built into ATM. At the local area level, ATM uses advanced encoding schemes to support 25-, 100-, and 155 Mbps on twisted-pair and multimode fiber media [4]. Similar bandwidths are supported in the wide area network (WAN).

Private networks have evolved over the past quarter century; successive architectures—all still in use—include X.25 and Frame Relay. ATM technology is capable of overcoming its antecedents' shortcomings in bandwidth, scalability, traffic-handling capabilities (also called *traffic management*), and suitability for combined data, image, and voice traffic.

ATM technology affords the following benefits, among others:

1. It supports high-speed communications (DS3, OC–3, and higher in the future).

2. It is a scalable technology, so that higher speeds can be supported without having to obsolete the underlying technology infrastructure.

---

[4] This chapter is synthesized from a number of sources published by the author, including but not limited to [1], [2], and [3].

3. It supports both WAN and LAN systems, enabling the user to deploy a single technology for all the enterprisewise needs.

4. It supports multimedia communications.

5. It supports multipoint communications.

6. It supports quality-of-service contracts.

7. It can be used as a multiservice platform to support a variety of legacy services, such as LAN bridging, LANE, MPOA, IP switching, frame relay, and circuit emulation.

In addition, the technology is cost-effective (because of the efficiencies achieved by overbooking of user bandwidth), reliable, and well-supported by the industry.

In the narrow sense, ATM is a cell-based *Data Link Layer* (DLL) protocol. A DLL protocol has the job of moving information across a single link in a reliable manner, using well-defined frames (here, cells). In this view, ATM only fulfills the lower portion of the data link layer. The DLL protocol relies on a *Physical Layer* (PHY) protocol, which can operate on any number of media, including DS3, OC-3, TAXI/FC multimode fiber, and UTP (CAT5). Protocolwise, ATM and the PHY taken together correspond to a MAC layer of a LAN (except that the former is connection-oriented at the DDL, while the latter is connectionless at the DLL). In general, there is a need to fill out the DDL before a network-layer protocol (e.g., IP or IPX) can be carried. This is done using a set of protocols called the *ATM Adaptation Layer* (AAL) protocols. LANE and MPOA also address these interworking issues. It should be further noted that the protocol model has three stacks: the User Plane, the Control Plane, and the Management Plane. ATM can be used in all three planes as the common link protocol, although the upper layers would be different.

In a more colloquial sense, ATM is the entire technology that supports fast-packet broadband communications. It is a statistical multiplexing technology, where the industry has agreed on how to support open multiplexing. With this technology, providers can overbook user bandwidth, thereby reducing network facilities; this, in turn, reduces provider costs, which, in the final analysis, can reduce the cost to the user.

Computing has for the past dozen or more years followed *Moore's law:* The density (and, hence, the potential performance) of silicon-based microprocessors doubles every 18 months. This is about an order-of-magnitude improvement every five years. This observation—named after its originator, Intel cofounder Gordon Moore—is governed by technology. By contrast, networking technology has advanced more slowly than computer technology, as measured by, say, user-available bandwidth. The author has shown elsewhere [5], in what might be called an equivalent law, that the bandwidth of WANs has increased by an order of magnitude every 20 years (this law is based on a retrospective analysis covering 60 years). Raw bandwidth growth has been governed principally by technological factors. However, the traditional gap between computers and networks (both LANs and WANs) has also been influenced by the lack of and ineffectiveness of a complete set of communication standards, by their slow evolution, and by their often-cautious adoption. Embedded physical facilities of public networks, spread all over the world, represent a large investment. Telecommunications companies are accustomed to multidecade system lifetimes. Therefore, there is a tendency to retain embedded technology for a significant amount of time. As a result, ten-year-old networks are commonplace, and many are likely to remain in service for some time. Network software has, on the whole, advanced even more slowly than computer software.

Networks are also limited by the capacity of the transmission media through which data and other signals are transferred. These physical channels have evolved from analog twisted-pair wire, intended initially only for voice traffic, through more advanced uses of copper via digital modulation or baseband methods (e.g., Asymmetric Digital Subscriber Line [ADSL] or Symmetric Digital Subscriber Line [SDSL]), to today's fiberoptic facilities. Digital switches have gradually displaced electromechanical equipment. Wide-area ATM is driven by advances in fiberoptics as well as by high-speed switching. The capacity and overall cost-effectiveness of fiber are attractive and are giving some impetus to the transition. Fiber is already predominant for major long-distance routes in the United States and local access usage is on the rise.

Networking standards are needed to make it possible to connect systems of numerous manufacturers and to facilitate interoperability of networks having different specifications. The ATM standards were originally based on Broadband Integrated Services Digital Network (B-ISDN) standards, which in turn evolved from ISDN standards. For LAN and campus environments, a number of newly developed standards have also evolved in the recent past.

To reemphasize the point, ATM is usable in both LAN and WAN environments. As implied in the previous section, many private networks are also moving to improve backbone as well as desktop capacity. Recent developments are extending the practical lifetime of existing media in private networks:

- Introduction of switching into the traditional all-shared-media networks, such as Ethernet, Token Ring, and FDDI
- Adoption and near-standardization of compression techniques that now reduce the bandwidth demands of images and video, making 10Base-T and 100Base-T LANs at least marginally capable of supporting these media

For a while, at least, all these network architectures will continue to play roles in private networks. Increasingly, however, ATM will displace these architectures or be utilized as an intermediate layer supporting other protocols such as TCP/IP. ATM offers the following features:

- Supports both long-distance and local networks
- Accommodates a wide range of physical media
- Transports voice, data, images, and video
- Supports QoS capabilities, with concomitant congestion control
- Accommodates bursty traffic and avoids congestion
- Scales to much higher bandwidths than frame relay

An ATM platform can be used to support a variety of legacy services, such as LAN bridging, LAN emulation, frame relay, circuit emulation, and so forth. For WAN applications, the switch and/or customer premises equipment (CPE) concentrator provide interworking (protocol conversion) functions that enable the legacy frames (and/or bit streams) to be mapped into ATM cells. This is done by using appropriate AAL protocols.

Development of ATM technology officially started in January 1985, although trial work had started a few years earlier. By 1989, many of the standards required to support WAN data communications were available, and equipment started to appear. Additional standards and equipment have appeared ever since. For WAN data applications, ATM technology is well into the second generation of equipment, with a third generation expected by press time.

What makes the difference between ATM and the solutions from which it has evolved?

- ATM offers such architected features as quality of service, statistical multiplexing, and traffic loss prioritization.[5,6] ATM supports a connection-oriented service, rather than a connectionless service. With its guaranteed quality of service, it is better suited to carry video and multimedia.

[5] ATM per se does not support per-cell cell-treatment priority from a performance point of view. (There is a loss priority, but this is not directly related to expediting cells through the switch or rendering special treatment.) To compensate for this, implementers have developed an *external* mechanism of providing per-session *traffic* priority discrimination. In a Permanent Virtual Circuit environment, the cells belonging to a session are implicitly assigned a priority at the switch by: (1) making a notation of the specific port on which the connection terminates, (2) adding some supplementary cell-enveloping at the switch as the cell traverses the switch (this supplementary information is, however, removed as the cell exits the switch), and (3) routing cells to different buffer pools based on the supplementary information. This implicit mechanism allows different traffic flows to receive different switching treatment (e.g., constant bit rate, variable bit rate, etc.). For Switched Permanent Connections, the treatment of the block of cells associated with a connection is based on the kind of service class called for in the setup message sent by the user for that connection at the start of the session. Again, the switch will have to use extensions of ATM to service the different requirements; in addition, the priority is not on a per-cell basis. Hence, the issue of priorities in ATM has to be carefully worded: One can say that mechanisms exist to support a kind of connection-level grade of service that gives the appearance of supporting, from a performance point of view, a weak kind of traffic priority.

[6] One of the consequences of the previous note is that QoS can be implemented differently by different vendors, because this is related to how they manage their policing, output buffers, shared buffers, switch, and so forth. There are no specifications as to what exactly it means to implement variable bit rate services and so forth. Furthermore, there are no specifications on how a carrier using a certain vendor's switch implements overbooking on the backbone side of the network—again leading to possible differences in service quality.

- ATM's bandwidth scalability is capable of accommodating, with appropriate hardware upgrades, the expected growth in end-user bandwidth demand. This expected increase is being driven by the rapid increases in performance and output of the computers served by the networks and by the growth in numbers of users as desktop systems become almost universal. ATM's basic philosophy of scalability provides technological longevity.

- Desktop workstations based on the most powerful reduced instruction set (RISC) microprocessors can already overtax a FDDI connection when used for I/O-intensive functions. PCs will reach similar output in a generation or two. As desktop users have deployed applications like hypertext and hypermedia and Lotus Notes with appended voice and video, demand for network capacity has accelerated. ATM will be better able to support these and similar applications than existing communication services.

- There are increasing requirements for high-speed Internet[7] access, for the development of private intranets, for the establishment (on the part of an increasing number of providers) of high-speed Internet backbones, and for the support (by carriers) of a carrier-provided hybrid[8] intranet. All of these evolving enterprise network combinations require broadband network services, such as ATM.

- Bandwidth must be allocated to users and applications equitably, on demand, and efficiently. This feature cannot be added as an afterthought, but must be part of the original architecture—one of ATM's strongest features.

- In addition to commoditizing the physical layers, the simplified infrastructure made possible by ATM's basic philosophy permits a much longer productive life cycle for the cabling plant and networking hardware than for other LAN technologies. The cost and

---

[7] We simply view the Internet as a collection of independent backbones by various providers, which have publicized or unpublicized meet points, for the support of inter-enterprise communication.

[8] These are intranets built over the public network (e.g., HTTP/HTML servers in the carrier's central office), but with enough closed user-group capabilities and security features to appear to be private to the corporation.

inconvenience of performance upgrades are reduced. For example, a change from Ethernet to fast Ethernet requires a replacement of adapters and other hardware. The ATM/SONET interface permits increases in speed with fewer, less costly adjustments. However, it must be understood that if the user purchases a 51.84-Mbps card, it will likely not support 155 Mbps or 622 Mbps; a new card would be required. The advantage of this approach is that the entry-level price is small, in that the user is not buying features which are not immediately needed. On the other hand, the user may choose to purchase a more powerful (but more expensive) card that can, by design, support 51.84, 155, and 622 rates immediately. The user can initially use this card only at the lower rate. As needs increase, the user can then upgrade the speed without any hardware changes. It should be clear, though, that a tradeoff is required: Nothing is ever free.

- ATM strikes a balance between two related issues in network design: accommodation of traffic bursts and control of congestion caused by competition for resources among candidates for transmission, particularly in the presence of bursts and overbooking.[9] Ethernet in the LAN and frame relay in the WAN cope with surges in network traffic, but both can become congested if too many users on the network want to transmit at the same time. Token Ring has a monitor that controls access, but it can backlog computers for unacceptable lengths of time, making them wait until a token lets them on the network.[10] ATM, in contrast, supports a form of bandwidth on demand that makes bandwidth available up to the maximum speed of the access line or switch. Altogether, ATM can deal more easily with bursts. It should be noted, however, that ATM, like anything else, has a limited bandwidth equal (at the switch level) to the bus speed—say, 2 Gbps. Also, there is a

---

[9] The much talked about traffic management issue only occurs in the case of overbooking—if no overbooking is instituted *anywhere* in the network, the need for traffic management virtually disappears. Naturally, this is tough to accomplish in the case of bursty traffic, unless the service is provisioned end-to-end based on the peak cell rate value.

[10] However, in Token Ring the delay is deterministic and bounded, while in Ethernet it is stochastic and (theoretically) unbounded.

certain maximum number of buffers—say, 64,000 per line card, or some larger number of shared-memory buffers. As soon as the user puts in more than a certain well-calculable number of inputs and arrivals, the user can become dead in the water, just like in frame relay and Ethernet. In addition, there are constraints based on both the access speed of the link (user or line side) and the speed of the trunk.

• Like any other technology operating at the data link layer, ATM is independent of upper-layer protocols (in principle, so are Ethernet, Token Ring, etc). This means that ATM will carry any PDU that is handed down to it through the syntax of the ATM service access point (SAP). From ATM's perspective, it is immaterial what is contained in the cell—IP, IPX, SNA, and so on. From the upper-layer protocol's point of view, however, there has to be protocol compatibility-matching: The protocol's PDUs must conform to the ATM's SAP: Hence, the upper-layer protocol must know how to hand off its PDUs to ATM. Having noted this tautological observation, there is a desire to retain existing upper-layer protocol applications unchanged. Hence, the developers have developed appropriate AALs to accommodate the interworking functions. AALs reside above the ATM layer and below the network layer. Note that ATM is a connection-oriented technology. This means that connections must be established at the ATM layer before information can be exchanged. There is also a desire to develop VLANs, where the logical community definition is independent of the physical location of the user. A number of products already exist to support VLANs, but these are vendor-specific. Many hope that ATM (in particular, LANE and MPOA technology) will become a vehicle to deliver vendor-independent VLANs.

• In ATM, the view of the network is nearly the same whether it is a LAN or a WAN, public or private. Thus, although ATM is an evolutionary step forward from earlier networking technologies, it is a major step.

• Support of legacy networks, as per the sections that follow.

This topic is discussed further in Chapter 3.

LANE

ATM technology must support existing LANs.[11] The ATM Forum's LANE specification defines how existing applications can operate unchanged over ATM networks. It also specifies how to communicate between an ATM internetwork and Ethernet, FDDI, and Token Ring LANs.

In this bridging interworking environment, Ethernet and Token Ring frames can transit an ATM network (in a segmented fashion) and be delivered transparently to a similar legacy network on the receiving end. Furthermore, a user on an ATM device can send information to an Ethernet or Token Ring device. LAN emulation provides users with a migration path from existing architectures without passing through successive stages of large-scale reinvestment (other migrations discussed in following sections use MPOA or IP switching).

To dispel some misconceptions about "ease of transition," however, it must be noted that the prospective user has to make an immediate investment to acquire LANE technology in order to support some of the ATM functions at this time. The transition is then from Ethernet or Token Ring to LANE to ATM. Another approach would be to save the funds that would be invested in this partial migration to ATM in order to be better equipped to make the direct migration from Ethernet or Token Ring to ATM at a later date. The LANE solution requires investments just to support today's relatively low bandwidth requirements; however, later it could be difficult to scale up to meeting the challenges of evolving network traffic patterns. Traffic concentration using ATM's higher capacity could overload tributary Token Ring and Ethernet networks by dumping bursts of information that would choke the receiving network. The desire to protect investments must be traded off against the risks of poor overall performance. In addition, solutions must be found for problems of addressing brought about by connections with legacy networks.

Specifically, LANE is an ATM-based internetwork technology that enables ATM-connected end stations to establish MAC-layer connections. It allows existing LAN protocols, such as Novell NetWare, Microsoft

---

[11] This section is based on promotional material from Fore Systems, Pittsburgh, Pa. Used with permission.

Windows, DECnet, TCP/IP, MacTCP, or AppleTalk, to operate over ATM networks without requiring modifications to the application itself. LANE provides the following capabilities:

- Data encapsulation and transmission
- Address resolution
- Multicast group management

The main components of LANE (see Figure 1.6) are as follows:

- The LANE driver within each end station (e.g., the host, server, or LAN access device)
- One or more LANE services (realized via specialized servers) residing in the ATM network

The LANE driver within each end station provides an IEEE 802 MAC-layer interface that is transparent to higher layer protocols, such as IP or IPX. Within the end station, the LANE driver also translates 802 MAC-layer addresses into ATM addresses, using an address resolution service (ARS) provided by a LANE server. It establishes point-to-point ATM switched virtual circuit (SVC) connections to other LANE drivers and delivers data to other LANE end stations.

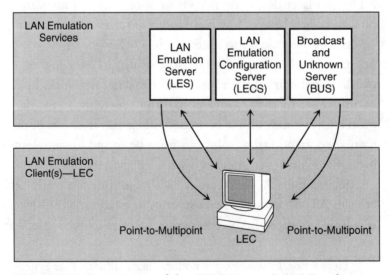

Figure 1.6 Components of the ATM Forum LAN Emulation specification.

LANE drivers are also supported on access devices (e.g., routers, hubs, and LAN switches) attached to the ATM internetwork. The access devices differ from end stations on the ATM internetwork in that access devices act as a *proxy* for end stations. As such, they must receive all multicast and broadcast packets destined for end stations located on attached LAN segments.

This combination of LANE drivers and services transparently supports the operation of existing 802.x LAN applications over the ATM internetwork. LANE is a logical service of the ATM internetwork. By using multiple LANE services, multiple 802 LANs can be emulated on a single physical ATM internetwork. This allows LAN administrators to create VLANs—logical associations of users sharing a common broadcast domain. These are also called *emulated LANs* (ELANs) in this context.

The advantages provided by LANE compare favorably to those of LAN bridging. LAN bridging technology was developed to support the expansion of local area networks. Ethernet bridges are transparent and require minimal configuration. Attached PCs do not require any modifications to operate in a bridged environment, saving much of the administrative cost associated with other internetworking technologies. LANE and classical bridging both support MAC-layer connectivity between LAN applications. LANE, however, removes the limitations of classical bridging, making it a key component of ATM internetworking.

> *Broadcast traffic reduction.*   Traditional LAN bridges have limitations that an ATM internetwork removes. LAN bridges do not scale to support very large networks. The flooding of unknown packets and broadcast traffic consumes valuable wide area bandwidth, leading to congestion on bridged LAN segments. In addition, bridges do not support active mesh topologies, limiting network performance.
>
> *Multiple VLAN support.*   A single ATM network with LANE supports multiple VLANs. Because each VLAN is distinct from the others, broadcast traffic in one VLAN is never seen in any other VLAN. It does not require any filtering or other mechanisms on stations not in that particular VLAN.
>
> *Dynamic configuration support.*   A LAN emulation configuration service (LECS) allows dynamic configuration capabilities within the ATM internetwork—eliminating the need to define the physical

connection between a host computer and the VLAN(s) to which it belongs. This allows a host computer to be moved from one building to another while remaining a member of the same VLAN.

Network security improvement.   The LECS also provides security and efficient bandwidth management. The system administrator controls VLAN membership, allowing the administrator to limit access to a particular VLAN.

Bandwidth management.   Users who commonly communicate with one another can also be grouped into the same VLAN. Another possibility is grouping users by specific types of traffic, such as IP or IPX. Therefore, managing membership to a VLAN based on the frequency of communication or traffic type uses bandwidth more efficiently.

Existing LAN support.   By using existing 802.x frame types and emulating the behavior of 802.x LANs, ATM network adapters appear to end stations and upper-layer protocols to be Ethernet or Token Ring cards—or both. Any existing protocol that has been defined to operate over Ethernet or Token Ring LANs can also operate over ATM LANE without modification.

ATM Forum Standard LAN Emulation

In 1995, the LAN Emulation subgroup of the ATM Technical Forum passed the LAN Emulation Version 1.0 specification. It defines the LAN Emulation User-to-Network Interface (LUNI) over which existing LAN protocols operate. The LUNI describes how an end station communicates with the ATM internetwork. Further work by the ATM Forum will focus on the LAN Emulation Network-to-Network Interface (LENNI). The Version 2.0 specification, providing enhancements in the area of redundant servers and Sub-Network Access Protocol/Logical Link Control (SNAP/LLC) encapsulation (per RFC 1483), is nearing completion at press time. LANE Version 2.0 begins to distinguish the elements within the LANE service cloud. It will accommodate multiple LES and BUS pairs by defining protocols between them. These protocols will provide a high level of scalability for LANE, and will support server function redundancy for improved robustness. There are also extensions related to QoS. QoS is designed to manage integrated voice, video, and data traffic in an ATM network. Through the use of different virtual connections, QoS supports applications that require constant, variable, available, and unspecified bandwidth. ATM switches can build a virtual circuit for each application and use QoS information to set up traffic priorities, choose network routes, and manage trunk availability. The degree of backward compatibility between LANE 1.0 and LANE 2.0 is being discussed at press time.

The ATM Forum LAN Emulation Version 1.0 specification consists of two components:

- LAN emulation clients (LECs)
- LAN emulation services, including the *LAN Emulation Server* (LES), the *Broadcast and Unknown Server* (BUS), and the *LAN Emulation Configuration Server* (LECS)

LANE services can be implemented on an ATM intermediate system; on an end station, such as a bridge, router, or dedicated workstation; or on a PC. They may also be implemented on ATM switches or other ATM specific devices. LANE services exist as a single centralized service where the LECS, LES, and BUS are implemented on an end station or ATM switch. But they can also be implemented in a distributed manner, where several servers operate in parallel and provide redundancy and error recovery. LANE services can operate on one or more LEC. For example, the LECS may reside on one end station, which is also a LEC, while the LES and BUS reside on another end station running LEC code.

LAN Emulation User-to-Network Interface

The definition of the LUNI model allows independent vendors to implement LANE end stations, while providing interoperability between their products. The LUNI defines initialization, registration, address resolution, and data transfer procedures for the interaction of the LEC and the LANE Services.

As the ATM Forum continues to develop LANE standards, the current specification allows for a wide range of implementations in the host computer. Pre-LENNI solutions are possible today and are available from a single vendor or multiple vendors who use the same signaling technique for network-to-network interfaces.

The LUNI defines initialization and registration. The LECS (server) controls the assignment of individual LECs (clients) to VLANs, using information contained in the LECS' database as well as information provided by each LEC.

Some observers believe that LANE will be difficult to implement as a network discipline across an enterprise because it imposes a large broadcast domain (a single IP subnetwork). Otherwise, if multiple ELANs are utilized, routers are needed to interconnect them; here routers can become bottlenecks. This topic is discussed further in Chapter 4.

MPOA

MPOA can be viewed as solving the problems of establishing connections between pairs of hosts that cross administrative domains, and enabling applications to make use of a network's ability to provide guaranteed quality of service [6,7]. For some time, already, manufacturers have released products that separate switching from routing, and allow applications to designate their required quality of service.

The MPOA working group of the ATM Forum is chartered with developing a standard approach to forwarding Layer 3 protocols, such as IP or Novell's IPX, transparently over ATM backbones. Building upon LANE, MPOA allows ATM backbone to support legacy Layer 3 protocols and their applications. When finished, MPOA will also allow newer Layer 3 protocols and their applications, such as packetized video application using IP's RSVP, to take advantage of ATM's quality-of-service features over the same ATM backbone.

MPOA will also enable the separation of the route calculation function from the actual Layer 3 forwarding function. This provides three key benefits: (1) integration of intelligent VLANs, (2) cost-effective edge devices, and (3) an evolutionary path for clients from LANE to MPOA. In the MPOA architecture, routers retain all of their traditional functions so that they can be the default forwarder and continue to forward short flows as they do today. Routers also become what are commonly called *MPOA servers* or *route servers,* and supply all the Layer 3 forwarding information used by MPOA clients, which include ATM edge devices as well as ATM-attached hosts. Ultimately, this allows these MPOA clients to set up direct "cut-through" ATM connections between VLANs to forward long flows without having to always experience an extra router hop [6,7].

A design desiderata of MPOA is to ensure that both bridging and routing are preserved for legacy LANs and the VLAN topology in use. An MPOA network uses LANE for the bridging function. An emulated LAN's scope (that is, an ELAN) is a single Layer 3 subnet,[12] whereas MPOA is focused on intersubnet (IP subnet) connectivity. Using LANE within the MPOA specification provides a number of benefits to the user, including the fact that it allows backwards compatibility, as an MPOA network can be built with both MPOA clients and LANE clients. In fact, the default operation for an edge device can be LANE until it learns more, simplifying the topology configuration and start-up operation.

For the Layer 3 forwarding function, MPOA is adopting and extending the Next Hop Routing Protocol (NHRP). This protocol is being defined by the IETF, and is expected to be published soon. NHRP is designed to operate with current Layer 3 routing protocols; thus, it does not require any replacement or changes to those protocols [6,7].

In order to set up a direct ATM connection between two ATM-attached hosts or between an ATM-attached host and an edge device, the ATM address of the exit point that corresponds to the respective Layer 3 address of the desired destination must be determined. An ATM-attached host can send an NHRP query to an MPOA server, that has been getting reachability information from routing protocols such as OSPF. The MPOA server may then respond with the ATM address of the exit point or ATM-attached host used to reach the destination Layer 3 address, or it may forward the query to other MPOA severs if it does not know the answer. Ultimately, the MPOA server that is serving the client that can reach the destination Layer 3 address will know the answer and reply. Once the replay arrives at the source, it can set up a direct cut-through ATM connection [6,7].

---

[12] As noted in the previous footnote, this implies that routers would be needed to connect ELANs.

In a VLAN environment, NHRP will reply with the ATM address of a router that serves the respective destination's VLAN. If this VLAN has more than one router connected to it, then there is no guarantee that the reply will address the router closest to the destination. When it does not, that router must bridge the data packets to the closest router, which will then forward them to the destination. MPOA is defining mechanisms in addition to NHRP which allow the MPOA servers to give out Layer 3 forwarding information to edge devices which represent the optimal exit point for a given destination so that this potential for excess hops is avoided.

This topic is discussed further in Chapter 5.

Network Layer Switching

The late 1990s have seen the emergence of various schemes to address several limitations of traditional IP routing, including:

1. The need for meshed or near-meshed physical networks
2. The requirement to perform Layer 3 processing at the endpoints of each link (in effect collapsing data forwarding and transmission, IP processing, and topology discovery into a single, obligatory function at each link endpoint
3. The relative complexity of Layer 3 processing
4. The duplication of Layer 2 and Layer 3 functionality (e.g., addressing)
5. The relatively poor use of improved data link layer technologies (e.g., ATM) by IP

To address these concerns, a number of vendor-specific as well as standards-based solutions have been proposed, are under development, or are being deployed. Notable vendor-specific solutions include Cisco's NetFlow and Tag switching technologies and Ipsilon's IP Switching technology. Standards-based solutions include MPOA, already discussed, and multiprotocol label switching (MPLS). Table 1.1 compares some of these Layer 3 switching technologies.

The ability to switch data based on very fast hardware table lookups on the MAC or ATM addresses can produce very fast and reliable networks. However, these technologies also pose problems of scalability and complexity that are seen by large leading edge Internet service providers (ISPs) or in enterprise network design. Of the problems introduced by large Layer 2 networks, some of the more pressing concerns have to do with smoothly integrating Layer 2 switching with Layer 3 switching. There are many vocal opponents of the current protocols who claim that the ATM Forum and the IETF have taken radically different approaches and that most of the work already produced is entirely too complex for practical value. Of additional concern are scalability and ease of operation, which are critical when building large intranets. High degrees of scalability can be difficult to achieve with a Layer 2 switched network because the address space is nonhierarchical. The combination is difficult to find with a pure Layer 2 switched network, so network managers have traditionally used network layer protocols, like IP, to fill that void.

The goal of network layer switching is to provide new means for interworking Layer 2 and Layer 3 technologies. Where the interworking differs from previous protocols is that the functionality of traditional network layer protocols, such as IP, play a more important role in the overall control of the network. In a network layer switched environment, all of the ATM switches understand and are capable of routing IP packets using protocols like the Border Gateway Protocol (BGP), or Open Shortest Path First (OSPF). In this model, the benefits of ATM that are applied to network design are basically speed and traffic control. Few of the higher-layer

Table 1.1 Enterprise Multilayer Switching

TECHNOLOGY	BENEFIT	APPLICATION
NetFlow	Scales router performance	High-touch router enhancement
Distributed NetFlow switching (e.g., MPOA)	Inter-VLAN cut-through	Campus wiring closet or data center
Tag switching	Scales intranet and Internet	Enterprise MAN and WAN router backbones

Table used courtesy of Cisco Systems.

ATM functions, such as ATM Forum signaling (described in Chapter 4), are used.

Practical experience with ATM and IP integration has given network designers a different insight on the technology that deviates somewhat from ATM's initial goals. Most first-generation ISP or Enterprise ATM deployments utilized ATM for either its speed or its ability to provide strict controls over traffic flow. The high speeds come in the form of 155-Mbps optical interfaces, and traffic control is achieved via permanent virtual circuit assignment. While this use of ATM has been useful, it is a very coarse method for designing a network that can sometimes lead to unexpected traffic flows. Standards activity and recent network deployments have focused attention on issues surrounding traffic engineering and the ability to use the Layer 2 network to explicitly define the route data follows. Traffic engineering is one of the more pressing needs and continues to drive much of the work related to IP switching. Ultimately, the tools described will allow network managers to build and control traffic flow across any Layer 2 technology in a predictable manner.

The body of work in the field of network layer switching can be subdivided into two categories, based on the level of granularity that is applied when mapping IP traffic to ATM virtual circuits. These models are *flow-based* versus *topology-based*. From a high-level view of network switching, one can think of the flow-based models as building a network out of routers or switching devices in which unique ATM virtual circuits are created for each IP *conversion,* where *conversations* are synonymous with a file transfer or WWW session. In the topology proposals, the routers or switching devices use their ATM fabrics to create ATM virtual circuits that can carry all of the traffic destined between pairs of subnetworks or IP routes.

The goals of network layer switching are similar regardless of the exact technical solution, because network layer switching's fundamental motivations are to remove the excess computational processing done during packet transmission by dividing routing from forwarding and then removing routing from the process whenever possible. In both the flow-based and topology-based models the ATM switch must be aware of IP and be capable of participating in IP routing protocols. However, the processes of forwarding and routing still maintain a clear division. Once the control process (i.e., routing) has detected either a route or a flow, it removes itself from the communication path and employs high-speed forwarding from the ATM fabric. These various proposals are explored in detail in Chapter 6, along with descriptions of how to use them when building Layer 3 switched networks.

Cisco's NetFlow and Tag Switching Technology

Cisco has been the first major player to address the issue of Layer 3 cut-through, via its NetFlow and Tag switching technology. This technology supports flow-oriented switching for multiple protocols. The approach is to *learn once—switch many times*. Cisco has positioned Tag switching as a LAN technology and NetFlow as a WAN technology—fundamentally, they are similar in concept.

A *flow* is a unidirectional sequence of packets between a given source and destination. Issues related to flows are: "What do I use to define a Net-Flow?"; "What determines the start of a NetFlow?"; "What determines the end of a NetFlow?"; and "How does one time out a NetFlow entry?" NetFlow granularity can be defined in terms of application (application layer applications, such as Telnet, FTP, etc.), transport layer protocols (e.g., TCP and UDP), network layer IP parameters (e.g., IP address), and data link layer protocols (e.g., Ethernet and Token Ring).

The IP header contains a protocol field (the 10th byte) that can be used to define a flow (e.g., the protocol could be ICMP, TCP, UDP, etc.). In turn, the UDP and TCP headers contain port numbers that define the nature of the data being carried (e.g., port 53 for DNS, port 520 for RIP, port 161 for SNMP, port 23 for Telnet, port 21 for FTP, and port 80 for WWW). Hence, NetFlow granularity can be defined at the TCP/UDP source or destination port, IP protocol type, and IP source or destination address. The NetFlow flow can start with a TCP SYN flag and terminate with a TCP FIN flag.

Figure 1.7 depicts how NetFlow switching works on a Cisco router. An incoming frame is first copied to packet memory. The top portion of the figure illustrates the kind of processing that is undertaken for incoming frames; the lower portion of the figure illustrates the various routing and switching tables that are maintained in the router.

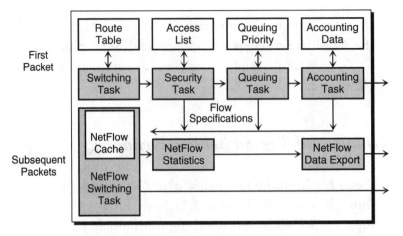

- Only first packet is processed by multiple tasks

- Connection-oriented NetFlow is defined with specific service requirements based on source/destination network address and transport layer port numbers

- Single switching task applies network services and collects traffic statistics

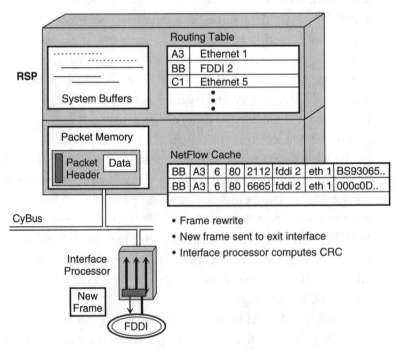

Figure 1.7 *Netflow switching technology (Cisco 7500 Series).*

When a new frame arrives, there may be no match in the NetFlow switching cache. Hence, the packet is copied to the system buffer. A lookup in the Layer 3 network address table is undertaken to see where the frame should be routed. The NetFlow switch cache is initialized, and the frame is sent to the exit interface. For the next frame, it is copied to packet memory, but a match is found in the NetFlow switching cache. This implies that the frame can now be sent directly to the output interface without having to go through the additional IP routing processing—specifically, frame deenveloping and routing table lookup, which can be relatively demanding in terms of resources.

Tag switching, also advanced by Cisco, is applicable at the campus network level. Tag switching addresses the throughput, scaling, and traffic engineering issues of corporate enterprise networks. It permits a graceful evolution of routing, and is intended to allow integration of ATM and IP. Tag switching combines Layer 3 routing with label-swapping forwarding (such as is available on Layer 2 ATM and frame relay networks). The simplicity of Layer 2 forwarding offers high performance; the separation of forwarding for long flows and routing aids the evolution of routing. Figures 1.8 and 1.9 depict the Tag switching operation.

Forwarding is based on a label-swapping mechanism, as well as on a control component that is used to maintain and distribute bindings. The router maintains a tag forwarding information base (TFIB), whose entries include the incoming tag and one or more subentries, such as outgoing tag, outgoing interface, and outgoing MAC address. TFIB is indexed by the incoming tag; TFIB may be per box or per incoming interface. The forwarding algorithm works as follows:

1. Extract the tag from the incoming frame.
2. Find the TFIB entry with the incoming tag equal to the tag on the frame.
3. Replace the tag in the frame with the outgoing tags.
4. Send the frame to the outgoing interface.

In working this way, the label-swapping mechanism is really like an ATM switch. Note that the forwarding algorithm is network layer independent. A tag distribution protocol (TPD) is used to distribute tag bindings to

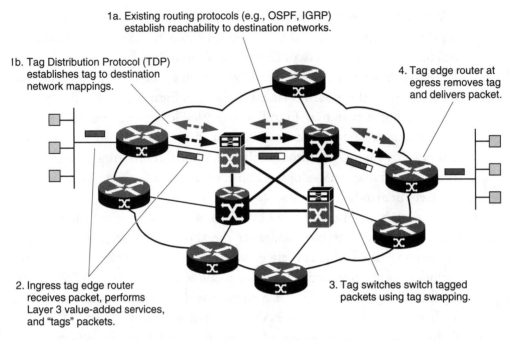

1a. Existing routing protocols (e.g., OSPF, IGRP) establish reachability to destination networks.

Ib. Tag Distribution Protocol (TDP) establishes tag to destination network mappings.

4. Tag edge router at egress removes tag and delivers packet.

2. Ingress tag edge router receives packet, performs Layer 3 value-added services, and "tags" packets.

3. Tag switches switch tagged packets using tag swapping.

Figure 1.8 *Tag switching operation.* (Courtesy of Cisco Systems.)

neighbors; the protocol only sends information if there is a change in the routing table and the device does not have a label.

IP Switching

IP Switching is a new networking technology advanced by Ipsilon Networks that combines the control of IP routing with ATMs speed, scalability, and quality of service to deliver millions of IP packets per second (PPS) throughput to intranet and Internet environments. IP Switching is quickly becoming the high-speed solution of choice for IP networks. Its evolutionary role is depicted in Figure 1.10. IP Switching implementations have already propagated across the entire spectrum of IP internetworking—from network interface cards to edge systems, telecommunications devices, and backbone, campus, and workgroup switches.

An IP switch implements the IP protocol stack directly onto ATM hardware, allowing the ATM switch fabric to operate as a high-performance link

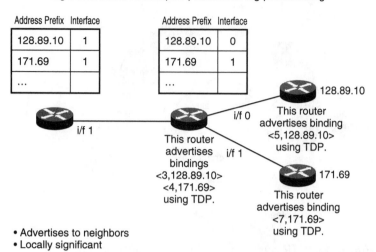

Tag Distribution Protocol (TDP) advertises tag-prefix binding.

Address Prefix	Interface
128.89.10	1
171.69	1
...	

Address Prefix	Interface
128.89.10	0
171.69	1
...	

i/f 1

i/f 0   128.89.10

This router advertises binding <5,128.89.10> using TDP.

This router advertises bindings <3,128.89.10> <4,171.69> using TDP.

i/f 1   171.69

This router advertises binding <7,171.69> using TDP.

• Advertises to neighbors
• Locally significant

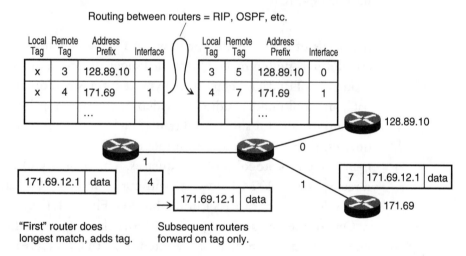

Routing between routers = RIP, OSPF, etc.

Local Tag	Remote Tag	Address Prefix	Interface
x	3	128.89.10	1
x	4	171.69	1
		...	

Local Tag	Remote Tag	Address Prefix	Interface
3	5	128.89.10	0
4	7	171.69	1
		...	

128.89.10

171.69.12.1	data

1

4

171.69.12.1	data

0

1

7	171.69.12.1	data

171.69

"First" router does longest match, adds tag.

Subsequent routers forward on tag only.

Figure 1.9 *Tag switching example.*

layer accelerator for IP routing. Based on IP, a well-understood and time-tested technology, an IP switch delivers ATM at wire speeds while maintaining compatibility with existing IP networks, applications, and network management tools.

Using intelligent IP Switching software, an IP switch dynamically shifts between store-and-forward routing and cut-through switching based on

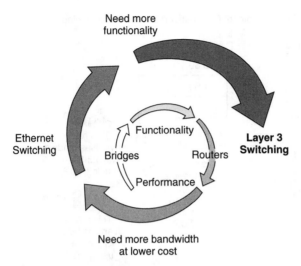

Figure 1.10 *IP Switching router functionality at switching speeds.*

the needs of the IP traffic, or flows. An IP switch automatically chooses cut-through switching for flows of longer duration, such as file transfer protocol (FTP) data, Telnet data, hypertext transmission protocol (HTTP) data, and multimedia audio and video. It reserves hop-by-hop store-and-forward routing for short-lived traffic, such as domain name server (DNS) queries, simple mail transfer protocol (SMTP) data, and simple network management protocol (SNMP) queries. The majority of data is switched directly by the ATM hardware, without additional IP router processing, achieving millions of PPS throughput (see Figure 1.11).

One of the key advantages that IP Switching offers is the interoperable network architecture that has resulted from the widespread acceptance of the technology. While router improvements continue to be limited to individual platforms, IP Switching can become a networkwide solution. IP Switching, with its published and widely implemented protocols, provides cooperative benefits to all peer participants—direct cut-through connections across the network and low latency compared to traditional router networks.

In an IP-switched network, all hosts and switches communicate through a common set of cooperative protocols—the Ipsilon Flow Management Protocol (IFMP; IETF RFC 1953, 1954) and the General Switch Manage-

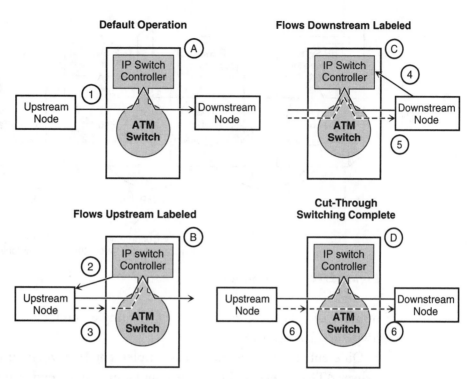

Figure 1.11 *An IP switch in operation: dynamic store-and-forward and cut-through switching.*

ment Protocol (GSMP; IETF RFC 1987)—to optimize short- and long-lived conversations between a sender and a receiver. Routing decisions need only be made once. As soon as longer-lasting flow data has been identified and cut through, there is no need to reassemble its ATM cells into IP packets at intermediate switch points. Thus, traffic incurs minimal latency, and throughput remains optimized throughout the IP-switched network. End-to-end cooperation also enables true QoS implementation, since with cut-through switching, policies naturally span administrative boundaries (see Figure 1.12).

Multivendor support for IP Switching also has resulted in a new economic model for high-performance networks. Customers can select the best platform according to their particular needs in a range of different network environments—from the campus, to the Internet, to the carrier networks, even at home. They can choose from a broad array of IP-based supporting applications, platforms, and tools. And they can work with any

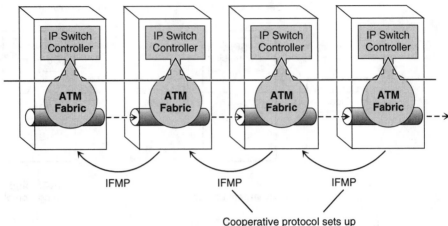

Figure 1.12 *The IP-switched network.*

number of solution providers, since no single vendor dominates the IP Switching landscape.

Different network elements can implement IP Switching protocols to deliver ATM-accelerated IP while maintaining multivendor, multiplatform interoperability. There is broad applicability for IP Switching across workgroup, campus backbone, Internet, WAN, and broadband access environments. Early applications included the following:

- IP Switching in the campus backbone
- IP Switching in workgroups
- IP Switching in the Internet and across the WAN
- Broadband access to IP-switched networks

Early product entrants include Advanced Telecommunications Modules, Ltd. (ATML), Digital Equipment Corporation (Digital), Efficient Networks, Ericsson, FORE Systems, General DataComm (GDC), Hitachi Telecom (USA), Ipsilon Networks, NEC America, Network General, RADCOM, and StarBurst Communications.

IP Switching in the Campus Backbone

In this scenario, IP Switching platforms deliver 155-Mbps throughput to campus backbones. Consolidation of enterprise resources in the data center, collaborative computing across workgroups, Internet and World Wide Web access, mobile computing, and multicast applications are among the many reasons that campus backbones are quickly losing power. IP Switching solves backbone throughput problems by delivering the speed and performance of ATM while maintaining full compatibility with existing IP hosts and applications.

ATM servers can connect directly to the IP switches to eliminate server bottlenecks, while IP gateways implementing the IP Switching protocols provide LAN connections to existing IP-based networks. Gateway software converts incoming IP-based Ethernet, fast Ethernet, or FDDI packets into ATM cells and forwards the traffic across the backbone.

IP Switching in Workgroups

In this scenario, ATM switch platforms from multiple vendors use the IP Switching protocols to provide dedicated 25- and 155-Mbps IP throughput to directly attached high-end servers, power users, and IP workgroups. Bandwidth-intensive collaborative computing software, multimedia applications, and large file transfers now have the throughput they need to operate effectively.

End users request IP file transfers and resources through standard IP transactions, and the data flow is accelerated throughout the network using the IP Switching peer-to-peer protocols. The result is on-demand high bandwidth, just the boost that is needed for optimal workgroup productivity.

IP Switching in the Internet and across the WAN

This scenario shows how carrier-class ATM switches can use the IP Switching protocols to accelerate IP data performance throughout the Internet and across the WAN. This underscores the native scalability of the IP Switching architecture. Because IP Switching distinguishes the data plane from the control plane, it can be implemented as easily in a carrier-class switch as in an enterprise or workgroup switch. And IP Switching can coexist side by side with ATM Forum protocols to allow carriers to take full advantage of standard ATM services as needed. This scenario also highlights IP switch connectivity over a private wide area network.

Broadband Access to IP Switched Networks

This scenario highlights the role of IP Switching in public network access. It shows how home consumers can use SDSL technology and IP switches to gain broadband access to the Internet and other IP-switched networks. With an increasing number of PCs at home being equipped with TCP/IP stacks, efficient TCP/IP support must migrate into access products of various types and complexities, ranging from xDSL multiplexers to central office telephony switches. The combination of IP Switching and SDSL technology in this scenario gives new meaning to high-speed networking for consumers.

IP Switching Applications and Network Analyzers

IP multicast is one of the hottest and most demanding applications emerging today. With IP multicast, a continuous data stream from a host can be transmitted to a designated group of users simultaneously from a single address. The interaction of this multicast address with standard IP multicasting protocols enables the network to replicate multicast packets at the most optimal locations, thereby conserving network bandwidth and resources. IP switches replicate multicast packets in ATM hardware, which results in a highly efficient, highly scalable implementation.

The topic of Layer 3 switching is revisited at length in Chapter 6.

Fibre Channel Standard

This section aims to discuss Fiber Channel technology. It does so by taking a more comprehensive look at channel technology than would be strictly needed to cover the immediate topic at hand. The purpose of the more comprehensive treatment is to expose planners to some of the issues related to broadband technology at the fundamental input/output (I/O) level. Fibre Channel has evolved from the need to provide higher-speed processor-to-peripheral connectivity, which is described in its historical context. It is also applicable at the LAN and WAN level. This discussion also helps the reader understand the key reason why FC was developed: FC evolved outward from the channel to the LAN. ATM evolved inward from the WAN to the LAN. Readers not interested in the background of FC may skip directly to the "I/O and LAN Technology" section.

Motivations

As microprocessors become more powerful, mass storage capacities requirements grow. For example, RAID (redundant array of inexpensive drives) systems are being increasingly deployed, particularly for multimedia and digital video applications. High-capacity optical discs are also seeing major deployment, as well as seeing evolution to more sophisticated technologies and to engineering methods supporting higher information packing density. Similarly, network operating systems and multitasking desktop operating systems demand more I/O power to access data, firmware, middleware, and other software modules. It follows as an immediate corollary of these observations that *I/O systems,* also called *I/O channels* or simply *channels,*[13] must evolve to support the bandwidth requirements for existing and new data-intensive applications. New peripherals also need to be supported. Desktop PCs, workstations, and servers supporting high-speed data, graphics, and video-based applications all require improved data transfer performance. In spite of all the hype about ATM and the information superhighways, no significant transition in this direction can be expected in the *whole* if PCs, workstations, and servers are not able to access the information from various silicon-based, magnetic, or optical storage systems at appropriate high throughput rates.

Applications such as high-speed data, video capture and retrieval, video dialtone, video on demand, multimedia, medical and scientific imaging, visualization, animation, virtual reality, modeling, and computer-aided design/computer-aided manufacturing (CAD/CAM), among others, require fast access to large files, whether in isochronous (real-time streaming) mode or in an asynchronous (non-real-time bulk download) mode. In order for these applications to blossom and enter the corporate and institutional mainstream, the microprocessor, the storage system, and the I/O subsystem all must handle large amounts of information over small intervals of time. Even more

---

[13] They are also referred to as *interfaces.*

mundane applications, including network and file servers, now require significant bulk memory-to-processor throughput.

In many situations, task bottlenecks are I/O- rather than computing-bound. Microprocessors operating at 50 million instructions per second (MIPS) are common today, and soon processors with capabilities of hundreds of MIPS will be a commodity item. According to *Amdahl's law,* a megabit of I/O capability is needed for every MIPS of processor performance. Some computing applications already require 1000 MIPS. This implies that the I/O requirements are in the 6- to 125-MBps range at this time, and will be higher by the end of the decade.

As a consequence, there is keen engineering interest in existing and emerging high-speed *storage* and *peripheral interfaces.* These include Small Computer System Interface (SCSI; the current version is SCSI-2), SCSI-3, Serial Storage Architecture (SSA), Fibre Channel (FC), Integrated Drive Electronics/AT Attachment (IDE/ATA), ATA Packet Interface (ATAPI), and Personal Computer Memory Card International Association (PCMCIA).

Buses and Channels

A computer system consists of a number of distinct components: processors, memories, and peripherals. These components must share and exchange information, consisting of either instructions or data. In order to do this, a physical means of transferring the information has to be provided, and a mutually accepted set of rules governing the transfers has to be established. This combination of protocols and physical medium constitutes a *datapath.* Two key datapaths of a computer are the bus and the channel. The *bus* normally interconnects computer components together, such as the central processing unit (CPU), the arithmetic and logical unit (ALU), the memory, the cache memory, and the I/O processor. The *channel* normally connects the computer with external peripherals, such as disk and tape drives, printers, and communication devices. Figure 1.13 illustrates a typical arrangement. Typically, the bus speed is higher than the channel speed (100 MBps compared to 10 MBps). In some computer systems the bus and the channel are the same. Buses, channels, LANs, metropolitan area networks (MANs), and WANs are all components on the continuum of technology used to interconnect computers and computing equipment over a wide range of distances and speeds. Each of these technologies has a characteristic but not rigidly defined set of features, as shown in Table 1.2 [8,9]. Buses tend to be viewed as connecting hardware cards and submodules to form a computer system, while channels, LANs, and WANs interconnect systems. However, the distinction between buses, channels, and LANs is becoming more blurred as time goes on. Figure 1.14 provides a view of various speeds of many of the important systems under discussion.

Drivers for New I/O Technology

Throughput, cabling issues, connector footprints, data integrity, and the need to support simultaneous I/O requests are all driving the introduction of new channel technologies in support of high-speed peripheral access

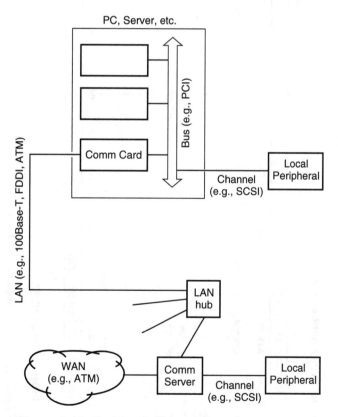

Figure 1.13 *Positions of buses, channels, LANs, and WANs.*

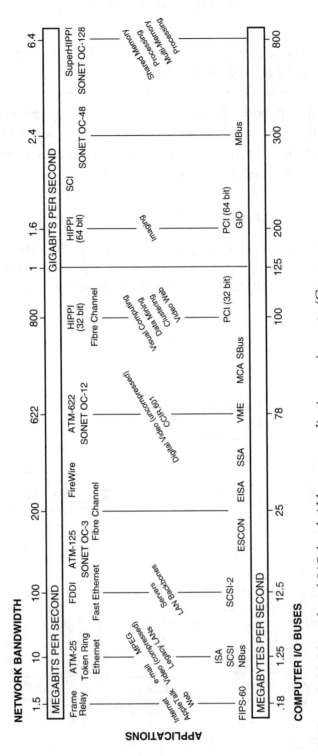

Figure 1.14 *Network and I/O bandwidth versus application requirements.* (Courtesy of Essential Communications, Albuquerque, New Mexico.)

44

Table 1.2 Comparison between Channels and Other Computer Networks

FEATURE	TYPICAL ENVIRONMENT			
	BUS	CHANNEL	LAN	MAN/WAN
Distance covered	A few tens of feet in a backplane or among cabinets	200–400 feet within a building	A few miles within a building or campus	Hundreds or thousands of miles
Type of devices interconnected	Parts of a computer system	Peripherals of a computer	Computer systems, servers, and workstations	Computer systems and terminals
Address granularity	Fine (a memory location)	Fine (peripherals)	Coarse (different computers and terminals)	Coarse (different computers and terminals)
Datarate	High (10–2400 Mbps)	High (10–1600 Mbps)	Medium (1–100 Mbps)	Low (1 Kbps to 1.5 Mbps and SONET/ATM rates are emerging)
Ownership of facility	Private	Private, but may also employ carriers' links	Private, but may also employ carriers' links	Typically carrier based
Cable	10–100 wires, parallel transmission	10–50 wires, parallel transmission	1 channel, serial transmission	1 channel, serial transmission
Topology	Bus	Bus	Bus, ring, star, or tree	Star or mesh
Error rate of cable	Very low	Low	Medium (depends on communications medium used)	Higher (depends on communications medium used)
Error detection	Usually not implemented	Implemented by channel extender hardware	Always implemented	Always implemented today; but perhaps less so with Frame Relay and ATM
Device relationship	Hierarchical	Hierarchical	Peer-to-peer	Master-slave
Response time	Media inherent	Very low (10–30 ms)	0.5–3 s	0.5–10 s
Data unit	Word	Byte	Message or file	Message or file

(mass storage in particular). With off-disk data rates climbing above 10 MBps, disk drives are now able to transfer data faster than the 8-bit fast synchronous 10 MBps SCSI buses they are connected to [10]. Table 1.3 provides a functional comparison of some of the important features of key channel interfaces used for such typical peripherals as disk drives, RAIDs, and CD-ROM jukeboxes.

Traditional I/O connections are parallel in nature, and use a relatively large number of wires (e.g., 50) to carry the various signals and bits. Evolving I/O systems such as Fibre Channel and Serial Storage Architecture are serial in nature, making the cable and the connector footprint much smaller. The demand for a higher performance, serial successor to SCSI is driven by the increasing need for greater connectivity for transaction processing. It is also driven by the need for handling the larger block size data transfer requirements of video and imaging applications on the user side, and by increasing disk drive performance on the peripheral side [11].

The benefits of the emerging serial interfaces are significant, but it may take until the turn of the century before they are adopted on a wide scale. Specifically, a current drawback of serial interfaces is that compatibility and interoperability at the software level is still being sought in the marketplace. Hence, it is unclear if one of these serial I/O technologies will emerge as an overall winner, dominating the late 1990s, or if a group of interfaces, including the parallel ones, will continue to play a role. The conservative perspective is that *all* of these interfaces will be important and will be deployed in some form or another in various classes of corporate and consumer equipment. Given that the product development cycle is now 12 to 15 months, and that each cycle actually represents a new product generation, it is conceivable that all the existing and emerging I/O interfaces will see actual commercialization before the end of the decade.

Table 1.3 Feature Comparison of Key I/O Interfaces

	SPEED	MAX DISTANCE	MAX DEVICES	COST/DEVICE	MATURITY	DEPLOYMENT
SCSI	40 MBps	25 m	15	<$15	14 years	>100 million
SSA	20 MBps	20 m	127	>$30	<2 years	<1000
FC-AL	100 MBps	100 m	126	>$30	<2 years	<1000

At press time, trade press like the following is being read:

> *What is the future of the current SCSI-2 interface? Will the next standard be Fast-20, a souped-up version of the current SCSI standard? Will it be a serial architecture like SSA, supported and promoted by its architect and chief sponsor, IBM? Will it be SSA's higher-performance arch-nemesis, Fibre Channel? Or, for the advocates of the electronic appliance, will it be (the IEEE) P1394 Serial Bus backed by such notable consumer electronic giants as Sony? [12]*

The issue, in the eyes of the entrenched providers, is: How does a manufacturer, much less an entire market, shift from the parallel I/O SCSI solution, which is widely proven and accepted, to a serial interface—an interface that is new, that is "completely alien" [10] to that which is already well understood, that has minimal market share as of 1997, and that has very few supplying sources?

Such discussions will probably occupy the arena of discourse for the rest of the decade, as proponents discuss the merits of each interface. The bottom line is that higher-performance I/O interfaces are invariably needed, and there will be a natural progression to the higher speeds and to the newer standards. But, as is the case with any new technology, the decision making is not only based on technical considerations, but also on such business realities as the installed base, hardware and software compatibility, external industry trends, and the competitive positioning taken by major computer manufacturers. Combined, the technology-centered arguments reduce to the rational considerations of "cheaper, better, and faster," while the business considerations are reduced to the pragmatic rules of competition and economics [12]. Migrating to serial requires a degree of change because the entire infrastructure of storage I/O is currently based on parallel technology; however, the transition is inevitable under the thrust of high-speed requirements and the continued drive toward the smaller and the mobile.

As the industry goes through transition, it is critical that accepted standards be employed in the development of new I/O systems in order to guarantee plug compatibility at the peripheral level, thereby, in turn, making connectivity easy and inexpensive to achieve. Table 1.4 identifies some of the key de jure standards applicable to this discussion.

Higher I/O speeds will drive requirements for higher LAN and WAN speeds.

Table 1.4 Key I/O Interface Standards

ATA-2
   ANSI BSR/X3.279: AT Attachment Extensions (ATA-2)
ATA-3
   ANSI BSR/X3 Project 2008-D: AT-Attachment-3 Interface
ATAPI
   ANSI BSR/X3 PN 1120D: ATA Packet Interface (ATAPI)
FCS
   ANSI BSR/X3 PN 1119D: Third Generation Fibre Channel Physical and Signaling Interface (FC-PH-3)
   ANSI BSR/X3 PN 1121 DT: Technical Report on SSA SCSI-2 Protocol (SSA-SCS)
   ANSI BSR/X3-P-1135-DT: Fibre Channel Copper Physical Interface Implementation Guide (FC-CU)
   ANSI BSR/X3.283: Information Technology—High Performance Parallel Interface Mapping to FC
   ANSI X3.230-1994: Fibre Channel Physical and Signaling Interface (FC-PH)
   ANSI X3.254-1994: Fibre Channel—Mapping to HIPPI-FP
   ANSI X3.271-Draft: Fibre Channel—Single-Byte Command Code Sets (SBCC)
   ANSI X3.272-Draft: Information Technology—Fibre Channel—Arbitrated Loop (FC-AL)
SCSI
   ANSI X3.T10/1071D (Fast-20)
   ANSI X3.131-1994
SSA
   ANSI BSR/X3T9.7/989-D: Serial Storage Interface PHY (SSA-PH)
   ANSI X3.PN1121DT: Serial Storage Architecture

Key Features

SCSI (SCSI-2) is a pervasive I/O interface. Currently, SCSI dominates intelligent parallel interface applications requiring multithreaded I/O. SCSI is now the primary choice for connecting workstations and servers to mass storage subsystems. However, in spite of its popularity and relatively high transfer rate, the system is under evolution to support higher speeds. In addition, competitive interfaces such as EIDE, SSA, and FC are making their presence felt. SCSI-3 is the standard for next-generation high-performance parallel I/O interface. Compared to SCSI and SCSI-2, SCSI-3 promises higher speeds, simpler interfacing, higher reliability, new device support (including RAID support), and new architectural approaches to device support.

SSA has been developed (by IBM) and accepted as a standard (by ANSI) to meet the performance requirements of systems ranging from PCs to mainframes. Fibre Channel is a multiprotocol serial interface covering a variety of computer environments, also extending to the LAN. It can be used in lieu of SCSI. FC technology is covered in detail in this book because it is a possible competitor to ATM as a campus technology.

The IEEE P1394 Serial Bus is a high-speed, global outboard device interconnect aimed at the consumer, industrial, and storage device markets. IDE is used extensively in the disk-drive context and on the AT attachment (ATA) bus. There is also interest in the attachment of CD-ROM and tape devices to the ATA bus, using the ATAPI protocol. Given that over 35 million ATAPI CD-ROMs were expected to be shipped in 1996, this is the second most common peripheral device interface. PCMCIA is the emerging memory and peripheral interface standard for a multitude of systems, including mobile and personal computing systems. In addition to the physical and electrical characteristics, card and system software are discussed.

SCSI

SCSI is currently the primary choice for connecting workstations and servers to mass storage subsystems. Figure 1.15 illustrates the basic arrangement. The standard was adopted in the 1980s and supported a transfer rate of 5 MBps on an 8-bit bus. A second generation of SCSI technology, known as *SCSI-2,* is now the "established standard for high performance" [13]. It utilizes a robust command set and supports a transfer speed of 10 MBps. In this book, *SCSI* refers to SCSI-2 unless otherwise noted.

Although SCSI provides a relatively high throughput rate, as more advanced and voluminous operating systems, such as Windows 95, are adapted, SCSI may become taxed when it is made to support both the volume and the datarate mandated by these systems. One approach employed by computer adapter manufacturers to extend the apparent throughput of SCSI-2 systems is the use of sophisticated caching algorithms; another approach, Fast-20, is discussed later, in this chapter.

As can be seen in Figure 1.14, computer systems are nearing the capacity to process I/O rates of over 100 MBps through the use of multiple processors and in-system I/O architectures like PCI [14]. In parallel to this,

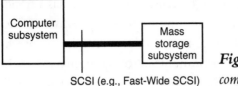

Figure 1.15 *Positioning SCSI in a computer/mass storage environment.*

and to overcome the performance limitations caused by disk access latencies, mass storage architectures are evolving from a single large disk to an array of disks, including RAID[14] and JBOD (just a bunch of disks). Given that a state-of-the-art disk drive can sustain a 10-MBps sequential data transfer, a disk array has the nominal capability to generate a transfer at 100-MBps. The byproduct of these advancements is that existing I/O communication standards used to support data transfer between the processors and the storage system, such as traditional SCSI, cannot keep pace with the performance required by the computer and the peripherals.

SCSI-2 was considered a high-performance interface until recently. However, RISC processors and systems such as Pentium have, in effect, brought the power of a mainframe to the desktop. In turn, this imposes a significant I/O requirement, particularly for emerging digital video and multimedia applications; this requirement is beginning to tax the capabilities of SCSI-2. Currently, the peak SCSI datarate is 20 MBps with 16-bit bus implementations (also known as *fast-wide SCSI*). The 8-bit implementation, which is common at this time (known as *fast SCSI*) has a maximum rate of 10 MBps. (The maximum bandwidth of SCSI is obtained by multiplying the data transfer rate by the width of the data path). To support increased throughput, system developers and integrators use multiple SCSI connections, as seen in Figure 1.16, where a number of fast-wide SCSI are utilized; or, worse yet, they employ proprietary channel and bus implementations. For example, given the preceding discussion, five fast-wide SCSI parallel channels are needed to support the transfer rate of contemporary PCI-based systems. The cost and complexity of this multibus approach may hinder further deployment along these lines. Hence, it is clear that SCSI has to evolve to remain a significant interface in light of the expected entry of high-performance ("bandwidth-hungry") applications.

---

[14] RAID technology development has been based on SCSI I/O.

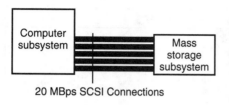

Figure 1.16 *Multiple fast-wide SCSI connections are required to support high-performance data rates (e.g., 100 MBps).*

20 MBps SCSI Connections

Table 1.5 provides a comparison of the various SCSI standards. For illustrative purposes, Figure 1.17 depicts a number of SCSI terminators and their prices at press time.

SCSI supports the following peripherals: Direct Access, Sequential Access, Printer, Write Once, and Processor. SCSI-2 supports all of the SCSI devices, plus CD-ROM, Scanner, Optical Memory, Medium Charger, and Communications. SCSI-3 supports all of the SCSI-2 devices, plus Graphic Arts prepress and Disk Array. A common way of adding SCSI to a PC is to install an ISA host adapter card in the PC. Depending upon whether it contains an onboard microprocessor, the SCSI adapter can cost anywhere from $100 to $300. A major portion of the cost is due to added components that increase intelligence. This added cost is justified because intelligence will minimize the performance impact of the slow data transfer speeds and long latencies of the ISA expansion bus. The new industry standard, PCI Local Bus, has the potential to reduce the cost of integrating a high-performance SCSI port by removing the drawbacks associated with older, slower expansion buses [15].

A brief description of RAID systems [16] follows, given the fact that SCSI is the key interface supporting this technology. Definitions have been established and standardized by the RAID Advisory Board for levels 0 through 5. RAID 2 and 4 are very seldom used since they are not considered practical solutions and are outperformed by RAID 3 and 5, respec-

Table 1.5 Comparison of Various SCSI Technologies

COMMON INTERFACE NAME (PARALLEL)	SPECIFICATION	THROUGHPUT ON 8-BIT BUS, MBPS	THROUGHPUT ON 16-BIT BUS, MBPS
SCSI	SCSI-1	5	
Fast SCSI	SCSI-2	10	
Fast-wide SCSI	SCSI-3		20
Double-speed SCSI	SCSI-3	20	40

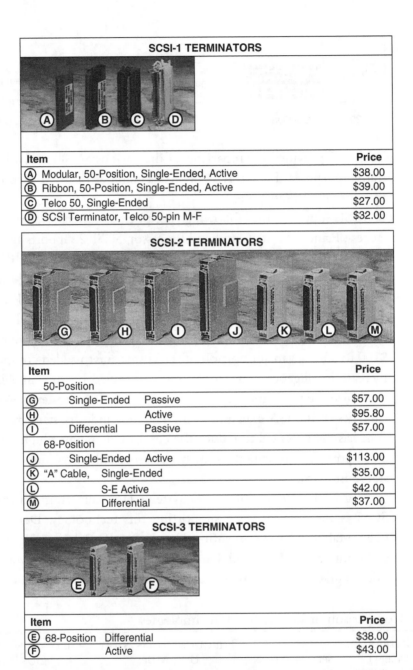

SCSI-1 TERMINATORS

Item	Price
Ⓐ Modular, 50-Position, Single-Ended, Active	$38.00
Ⓑ Ribbon, 50-Position, Single-Ended, Active	$39.00
Ⓒ Telco 50, Single-Ended	$27.00
Ⓓ SCSI Terminator, Telco 50-pin M-F	$32.00

SCSI-2 TERMINATORS

Item			Price
50-Position			
Ⓖ	Single-Ended	Passive	$57.00
Ⓗ		Active	$95.80
Ⓘ	Differential	Passive	$57.00
68-Position			
Ⓙ	Single-Ended	Active	$113.00
Ⓚ "A" Cable,	Single-Ended		$35.00
Ⓛ		S-E Active	$42.00
Ⓜ		Differential	$37.00

SCSI-3 TERMINATORS

Item		Price
Ⓔ 68-Position	Differential	$38.00
Ⓕ	Active	$43.00

Figure 1.17 *Examples of SCSI terminators.*

tively. Other RAID levels have been introduced by different vendors, but their definitions have not been standardized (for example, a RAID 6 configuration may refer to the utilization of two drives for redundancy, allowing two drives to fail before data is lost).

RAID 0. This is not, strictly speaking, a RAID system, since it does not provide any redundancy. It distributes data across all drives in the array in order to achieve performance. However, it can deliver higher performance only if tuned for specific applications and their I/O request characteristics.

RAID 1. This is traditional disk mirroring. RAID 1 may provide performance improvement on read requests, because the controller can read from whichever drive is closest to the requested data. The cost of mirroring is twice that of single drives (e.g., to store four drives' worth of data, eight drives are needed). Mirroring can be implemented in hardware or software.

RAID 0+1. This level combines disk mirroring with disk striping. Some vendors call this RAID 1 for marketing purposes because they have combined the features of RAID 0 with mirroring to achieve better reliability and performance.

RAID 3. User data is striped across a set of drives. A drive to hold parity data is added. Parity data (the combined binary value of the original striped data) is calculated dynamically as user data is written to other drives. Segments are set small (bits or bytes) with respect to the average request size. The parity drive must be accessed for every write, so write performance is slowed. RAID 3 is best suited for use when large sequential read requests are the primary array activity. They are a poor solution for random access and small data segments. The cost of RAID 3 is the cost of the data drives plus one parity drive and a controller.

RAID 5. With this level, user data is striped in large segment sizes and parity is computed and distributed evenly across all drives in the array. As a result, no single drive acts as a bottleneck to disk writes, as is the case in RAID 3. Usually, one drive participates in any request, allowing several requests to be handled simultaneously from all drives in the array. Since the segments are large, insufficient data is received to enable parity to be calculated solely from the incoming data stream, as in RAID 3. Incoming data must be combined with existing parity data to recompute the new parity. Thus, each write request involves reading from two drives (old data and old parity) and writing on two drives (new data and new parity). This results in poorer write performance than the other RAID levels. RAID 5 is most appropriate in environments that have small random I/O requests, such as the office, transaction processing, and decision support. The cost of RAID 5 includes the data drives plus one additional drive, since the parity is spread across all drives.

SCSI-2 has two management systems: (1) the *Command System,* along with Command Descriptor Blocks (CDBs) and their associated data and status; and (2) the *Message System,* which uses the Message In and Message Out phases. The command system is well defined in the SCSI-2 standard. Table 1.6 depicts the key SCSI commands. There is flexibility within this system to perform the various functions required by targets and initiators to meet customer needs. The message system, however, is less well defined in the SCSI-2 standard. This implies that there is the potential for incompatibilities between host bus adapters (HBAs as SCSI initiators) and targets (e.g., tape units, disk drives, etc.). The command and message systems are loosely coupled to manage and invoke the functions carried out by each logical unit in the targets [17].

Each message of the SCSI-2 Message System standard belongs to one of the following three categories:

Table 1.6 Key SCSI Commands

COMMAND	FUNCTION
Inquiry	Inquire about a device type, vendor, model, version, and feature.
Mode Select	Set device parameters.
Mode Sense	Get device parameters.
Read	Read data from a device.
Request Sense	Obtain sense data (error information) from a device.
Test Unit Ready	Determine if a device is ready.
Write	Write data to a device.

- *I/O process identification,* as part of an initial connection by a HBA or during reconnection by a target
- *I/O process management,* as part of an initial connection by a HBA or during reconnection by a target
- *Connection management,* related to the physical transfer of information between a HBA and a target

SCSI adapters operating at 10 MBps cost $120 to $250. Given low-cost high-performance competition from IDE controllers ($20 to $80 at 13 MBps), SCSI designs are being migrated to designs that include caching, in order to justify their higher cost. Caching can provide a significant increase in performance. A noncached PCI-to-SCSI controller may support a 10-MBps transfer; however, a well-designed cache-based SCSI controller with a tag-RAM architecture and 4 MB of onboard cache can support a transfer in the 30- to 33-MBps range [18].

Fast-20 SCSI

Recent parallel developments in SCSI from standards committees have succeeded in improving its throughput so as to meet the channel throughput requirements of multimedia applications; this near-term enhancement of SCSI allows a doubling of the data transfer rate to 40 MBps. This increase can better match the 100-MBps performance already required by high-performance computer and storage systems at this time, as discussed earlier. The advantage of this approach is that the systems can maintain the physical infrastructure, compatibility, and software environment of existing parallel SCSI implementations [14]. This faster version of SCSI doubles the transfer rate from 10 (as defined in SCSI-2) to 20 million transfers per second (MTPS). Hence, the transfer rate is 20 MBps on an 8-bit bus and 40 MBps on a 16-bit bus. This technology is variously known as *Fast-20, DoubleSpeed SCSI,* or *UltraSCSI.*[15]

Fast-20 is defined in a new SCSI-3 standards document, X3T10/ 1071D. It doubles the SCSI synchronous data transfer rate while maintaining compatibility with the current 5 and 10 MTPS synchronous rates as

---

[15] This faster version of SCSI was initially named UltraSCSI, but was renamed Fast-20 because it supports 20 million transfers per second.

well as with the SCSI asynchronous protocol. Fast-20 was originally conceived by Digital Equipment Corporation as a backplane-only upgrade for SCSI-2 systems. The concept has also been incorporated in the SCSI-3 parallel interface specification. Table 1.7 puts some of the features of SCSI (Fast-20) in context with the other interfaces.

The enhanced features offered by serial I/O solutions that are most attractive are also addressed by enhanced parallel SCSI, as follows [10]:

> *Connectivity.* SCA connector standards that specify a single connector for most I/O protocols are now available. This matches the serial connector size advantage, since products designed for emerging backplane-attachment applications will basically have the same connector regardless of which I/O protocol is used. In addition, since there are no cables in these applications, cabling is not an issue. For cabled environments, new 16-bit high-density connectors are now

Table 1.7 Comparisons of Key I/O Technologies

	SCSI FAST-20	SSA	FC ARBITRATED LOOP	P1394 SERIAL BUS
Max transfer rate, MBps	40	20 and 40, full duplex	100	25
Mode	Parallel, 8 and 16 bits	Serial	Serial	Serial
Max number of peripherals per bus	4–8	126–129	126	64
Cable type	Ribbon	Twisted-pair Fiberoptic	Coax Fiberoptic	Twisted-pair
Max length, m	1.5–3	40, twisted-pair 680, fiberoptic	30, coax 10,000, fiberoptic	4.5
Ports	1	2	2	1
SCSI-2 software/ firmware compatibility	Almost complete	Partial	Partial	Significant
Cost of link, compared to SCSI-2	Similar	Higher	Highest	Similar
Availability	1995	1995	1995	Unknown

available and are about the same size as serial connectors, while new quarter-inch 16-bit SCSI cables provide the implementor with nearly the same economy of space, airflow, and general user-friendliness promised by serial I/O cables.

Hot-plugging with automatic configuration. While parallel SCSI has been hot-plugging for years (if implemented with properly designed silicon with low leakage and high impedance capabilities), a new feature called SCAM has only recently been added. SCAM (SCSI Configured Auto-Magically) provides parallel SCSI with automatic configuration.

Increased transfer rates and bandwidth. The transfer rates achievable with Fast-20 (up to 40 MBps), should be adequate for the next two or three generations of hard-disk drives.

Fast-20 is the fastest standard I/O technology for computer-peripheral connections at press time, and has the property of maintaining *backward compatibility* to embedded SCSI implementations. For example, in Figure 1.16, only three Fast-20 (double-speed) buses would be needed to sustain the 100-MBps throughput. In addition, this design approach is easy to implement because it allows integrators and manufacturers to utilize existing cables, connectors, enclosures, and software. As a compatible extension of existing parallel SCSI implementations, Fast-20 effectively extends the useful life of parallel SCSI implementations until the new serial I/O solutions become commercially available. Existing disk drives and adapters within a system (e.g., a video server) can be replaced with Fast-20 SCSI disk drives and adapters to achieve higher throughput.

Given the fact that software and operating system (OS) drives for Fast-20 are backward compatible with existing hardware, users can utilize a single set of drivers for both the existing and the new devices. This is possible because SCSI has the ability to negotiate the transfer speed (e.g., 5, 10, 20, or 40 MBps); users can interconnect different devices on the same cable. Consequently, Fast-20 has been enthusiastically embraced as a life extender for the cable and backplane-based parallel SCSI architecture that is used extensively in storage subsystems, midrange systems, workstations, and high-performance PCs [12].

Fast-20 work has been undertaken by the X3T10-SCSI committee of ANSI. The committee includes representatives from the component, disk drive, host adapter, peripheral, and system manufacturers communities, all of whom are familiar with both the technology and the limitations of SCSI. Formal double-speed standards were under development at press time (a draft document for double-speed SCSI-3 was published in 1994 and is being used by some companies to develop component-level products). The specification is expected to become a full standard by press time. A number of disk manufacturers have made plans to employ this interface for high-performance drives at capacities of 1GB or higher. Many SCSI board-level manufacturers now offer SCSI-based RAID 0, 1, 3, 5, and 10 (striping plus mirroring) controllers. These allow a large number of hot-pluggable SCSI disks to offer both storage capacity and high availability for traditional and evolving mission-critical applications.

Fast-20, however, is seen by manufacturers of high-end systems and disk arrays as only an *interim* solution, because other limitations of SCSI I/O (beyond speed) have not been solved by the proposed upgrade. (Limitations include distance constraints, large connector footprints, and population of peripherals on a single cable.) Serial I/O designs such as FC and SSA solve the distance, footprint, and device limitations of SCSI. According to observers, by doubling the transfer rate to 20 MTPS (40 MBps throughput), the system should, based on conventional processor MIPS improvement rates, be viable up to 1997.

As noted, the principal advantages of Fast-20 are that it requires minimum redesign and is compatible with existing system software; the principal limitations relate to cabling considerations[16] and the number of permissible devices (this number is from 4 to 8—note that SCSI-2 permits longer cable lengths and 8 to 16 devices). In the short term, parallel SCSI enhancements provide a path to higher speeds via a transition that is compatible with the infrastructure in the existing marketplace. For instance, migrating to Fast-20 from non–Fast-20 8- or 16-bit SCSI entails little or no change to any of the following [9]:

- Cost
- Operating system I/O routines

---

[16] These are 2.5 meters for differential devices and 1.5 to 3 meters for single-ended devices (depending on the number of devices).

- Software drivers
- Controller and device firmware
- System and subsystem logical configuration (software)
- System and subsystem physical configuration (hardware)
- Cables and connectors
- System cooling
- System power distribution
- Documentation
- Test equipment
- Production equipment
- FCC qualification
- Brain trust (expertise, education, and personnel)

Fast-20 was designed from the beginning to be easily integrated into existing systems while supporting backward compatibility: Nearly all of the technical changes were made transparent to the implementor by placing the features in silicon. Once the silicon is upgraded, the implementor has only to consider the following simple changes for single-ended 8- or 16-bit applications [10]:

Active termination. For several years, SCSI channel problems have been eliminated by replacing the passive terminators with active ones (active termination has already become fairly standard within the industry). This has improved SCSI bus integrity, especially in 10–million transfer per second implementations.

Cables. Since the tolerance for Fast-20 cable impedance (84 to 96 Ω) places its range within the pre–Fast-20 impedance range (72 to 96 Ω), most existing 10–million transfer per second implementations utilizing quality cables will work fine.[17] Existing systems and platforms that currently implement 10-MTPS fast synchronous SCSI are the most likely to migrate to Fast-20 SCSI.

---

[17] It is generally desirable to use Fast-20 SCSI cable impedances even if the system is not running Fast-20 SCSI.

Length, load, and spacing. To prevent incompatible operation, the standards for both 10-MTPS and Fast-20 SCSI provide implementation guidance. A maximum cable length of 3 m with a maximum of 4 equally spaced devices, or a maximum cable length of 1.5 m with a maximum of 8 equally spaced devices is a conservative approach. This compares to the 3 m for 10-MTPS synchronous SCSI, with a number of devices which depend on the data-bit width of the cable. But there are implementations of 10-MTPS SCSI that successfully use 6 m of cable. Pragmatically, the recommendations for Fast-20 are nothing more than a definition of what also works best for non-Fast-20 SCSI implementations. Poor cabling and lumped loads will continue to be a problem.

SCSI-3

Work on SCSI-3 is underway. Industry observers are of the opinion that "parallel SCSI is supported by knowledge, experience and maturity; serial I/O interfaces will have to demonstrate that level of growth and maturity before it poses any real threat to the parallel SCSI marketplace" [19]. SCSI-3 adds major improvements to SCSI, making it a viable technology for the rest of the decade.

Enhanced IDE (EIDE) and Fast-ATA

The AT Attachment Interface, also known as ATA or IDE, has been used in the PC storage market for years. It was introduced by IBM in support of hard drive controllers in the early 1980s. It eventually evolved into Western Digital's IDE drive interface. Now it is undergoing additional adaptations to meet the evolving demands of another generation of equipment. At this time, enhanced derivatives of this disk peripheral interconnect, specifically Enhanced IDE (EIDE) and Fast ATA-2, are used on over 90 percent of the PCs being shipped; furthermore, this technology is included in about 70 percent of all discrete disk drives shipped at press time.

Original AT system BIOS were intended for disk drives with capacities up to 40MB. With each new generation of microprocessors, the need to handle larger amounts of data in a shorter amount of time has dictated significant interface improvements. IDE, introduced in 1984, provided a flexible interface solution that supported drives up to 512MB. However, as is now clear, the scope of traditional IDE is limited, as the demands for larger drives, faster transfers, and new peripherals become more pronounced.

During the late 1980s and early 1990s several standards committees, including ANSI, undertook work to expand the capability of the AT Attachment Interface in order to keep up with microprocessor and bus architectures. At the present juncture, processors such as Intel's Pentium and the PCI bus are used in juxtaposition with EIDE and Fast ATA-2 to support the kind of I/O that was originally found only in midrange and mainframe computers.

EIDE was introduced by Western Digital in 1994. The four problem areas of IDE that EIDE addresses are [19]:

- Support for disk drives larger than 528MB[18]
- Faster data transfer rates—EIDE can achieve 13.3 MBps, while IDE only supports 3.3 MBps
- A larger number of devices—EIDE's 4, versus IDE's 2
- More device types—EIDE's disk and CD-ROM drives, versus IDE's disk drive

EIDE- and Fast-ATA-based disk and controller technology is now replacing SCSI approaches for high-end desktop and low-end to midrange server applications. For example, PCI bus–to–IDE controller adapters, when properly designed, allow effective support of both high-speed IDE or ATA disk drives and slower ATAPI CD-ROMs or tape drives (see following); these can be connected to the processor through the same ATA interface cable.

The pursuit of higher IDE/ATA performance implies more stringent requirements for reliability and certification as the technology limits are pushed. Benchmarks now show that IDE/ATA drives support speeds of 13 MBps in sequential reads, which match or exceed today's fastest SCSI disks

---

[18] Standard AT is unable to support IDE drives with capacity exceeding 528MB because of the cylinder, head, and sector definitions of both BIOS Interrupt 13 and the IDE interface.

and adapters. Fast ATA-2 puts the emphasis on performance, improving SCSI-2's 10 MBps to about 13 MBps (by comparison, "low-end" IDE only supports 3.3 MBps). Fast ATA-2 does not require special host adapters, BIOS, or operating system changes: Conventional AT host adapters or motherboard connections are adequate. It should be noted, however, that there also are some drawbacks with Fast ATA-2, particularly as an alternative to SCSI-2 for RAID access. First, the ATA command set is limited, in that it lacks some performance capabilities, such as tagged command queuing. Thus it is not robust enough for some applications. Second, this interface only accommodates two devices per string (all over just 18 inches of cable), and there are no overlap commands. Third, the throughput may not be exactly as high as quoted since the speed is measured buffer-to-buffer rather than system-to-peripheral [13].

Newer IDE controllers support up to four drives on two cables or eight drives on four cables; these extended IDE features provide low-cost, high-performance alternatives to SCSI for hard disk, CD-ROM, and tape peripherals. However, EIDE is seen by some as having limitations for support of RAID, because it requires support beyond just the storage level. Specifically, to secure the benefits of EIDE, one must enhance the IDE BIOS, obtain an EIDE operating system, and have EIDE hardware implementations.

As an extension of the IDE technology, EIDE utilizes a dual-channel architecture (via dual IDE connectors) that provides the capability to support both a primary, high-speed (disk) channel and a secondary, slow-speed (nondisk) channel, as illustrated in Figure 1.18 [20]. EIDE supports disk capacities *exceeding* 528MB by a BIOS translation based on the logical block addressing (LBA) method or on the traditional cylinder, head, and sector (CHS) method.

EIDE is already bringing order to the CD-ROM market by offering an alternative to the proprietary methods used previously. Intel and Compaq have embedded EIDE within the PCI chipset, further strengthening its market share potential (proponents see a 100 percent market share).[19]

---

[19] SCSI CD-ROMs had a 2 million units penetration (as measured by shipments) in 1993, while IDE/EIDE had none; by 1995 each had about 6 million; from that point on, EIDE is expected to take over, with 14 million EIDE CD-ROM units shipped in 1997 versus 8 million with SCSI.

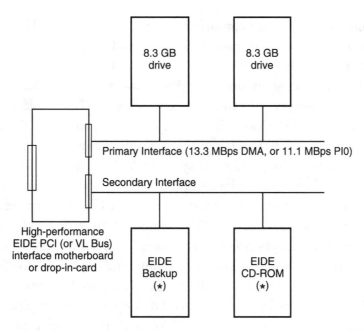

(*) or additional drive

Figure 1.18 *EIDE connections.*

EIDE has had good success as a desktop technology: Integrators and OEMs have not used SCSI extensively for single-user systems, based on cost considerations. However, at the server end, EIDE has not overcome SCSI because the factors that make EIDE appealing to a single user have worked against it in that context (e.g., limited peripheral attachment, lower-capacity drives, etc.).

Proponents expect to see EIDE "take over completely at the desktop level, and SCSI be left with a niche market consisting of scanners, color laser printers, and jukeboxes" [20] (some of these devices may soon be supported by ATAPI). However, SCSI is perceived to be a reliable, proven, high-performance technology, with a strong foothold in more sophisticated OSs, such as Windows 95 and OS/2 Warp, and in production servers, where I/O is most critical. Although SCSI technology will continue to be the principal approach for RAID (particularly in the 10- to 100GB range), some see IDE-based solutions becoming viable at the lower end and for RAID 1 configurations. Both hardware and software implementations are appearing at press time. Future IDE/ATA RAID implementations will include striping and striping plus mirroring (also referred to as *RAID 10*).

Recently, programmed I/O (PIO) has evolved to Mode 4 operations. This ANSI-approved extension decreases the ATA-bus data transfer cycle time from 240 ns (Mode 2) to 180 ns (Mode 3) to 120 ns (Mode 4). However, at these higher speeds, cables and subsystems may experience reliability challenges: Acquisition of an IDE/ATA drive controller combination should be based on vendor certification. Certification should also be sought across a range of system motherboards, BIOS, OS, and software applications.

Direct memory access (DMA)-based EIDE or Fast ATA-2 drives and bus master IDE controllers[20] are poised to become an industry standard, thereby diminishing, according to some, the importance of SCSI in multitasking, disk-intensive applications. New specifications for bus master IDE/ATA operation and faster hard disk operation were planned for 1996, creating even higher levels of IDE or ATA-type operations [18]. Parameter goals were: drives as large as 137GB; data transfer of 16.6 MBps in Mode 4 PIO or Mode 2 DMA; and support for disk, CD-ROM, tape, and CD-R.[21] The cable length will continue to be 18 inches. However, there are technical challenges in implementing Mode 3 and Mode 4 PIO insofar as the increased speed places tight quality tolerances on cables and connectors and may necessitate active components.

The ATA Packet Interface (ATAPI) standard for connecting IDE CD-ROMs devices is becoming a commonly accepted standard—eroding, in the proponent's eyes, SCSI's supremacy in the connection of non–hard disk devices. ATAPI was initially advocated by the Small Forms Factor Committee. It is now used by the majority of the PC manufacturers for connecting CD-ROMs, and in the next couple of years it will also be uti-

---

[20] There is an expectation that IDE manufacturers will migrate their chip-based controllers from pure programmed I/O transfer mode to SCSI-like bus master capability. The bus master logic (ASIC, to be exact) initiates its own transfer, offloading the PC's processor from this task. While the incremental cost of doing this is small, the performance compared to a SCSI-based solution is higher, further affecting the price-performance separation between IDE/ATA and SCSI approaches.

[21] Note that transfer rates as high as 17.5 MBps for sequential reads on a Pentium 90 using EIDE were already achieved in 1995—high-speed EIDE requires connector and cabling modifications and may require caching [20].

lized for tape support (ATAPI-based tape drives). To sustain reliable CD-ROM and tape I/O, the ATAPI interface includes several new commands to the ATA interface command sets.

SSA

IBM originally developed SSA and continues to be its champion. SSA was developed to address a gamut of hardware applications, including PCs, mainframes, and high-end storage devices. Serviceability and availability are major performance characteristics that were taken into account in the development of SSA. The physical and data link layers have now been incorporated by ANSI into the SCSI-3 standard.

Each SSA device supports a point-to-point link operating at 20 MBps. Full-duplex operation allows a link throughput of 40 MBps per device. Usage of a dual-port arrangement increases the throughput to 80 MBps. Additional improvements are planned at press time, aimed at doubling these speeds. Based on SSA's flexibility, devices can be configured in strings, loops, or switches. The distance covered is 40 m on twisted-pair (680 m on fiber), making the system viable for RAID and other large arrays. When SSA is used in a loop configuration it offers increased availability via multiple data paths and dual porting.

Instead of transferring data 8 or 16 bits at a time over a bus of the same number of parallel wires, as is the case in SCSI-based I/O interfaces, SSA transmits frames consisting of 6 to 139 characters of 10 serial bits each. These frames communicate data, protocol, or user-defined information. The SSA electronics are built on a single integrated CMOS chip. Hence, expectations are that these links will be low-cost. SSA drives are comparable to differential SCSI drives. In the future the drives should cost the same as a single-ended SCSI drive. However, it should be noted that use of SSA requires both device driver modifications and nontrivial changes in initiator code. Another consideration is that FC offers at least a fivefold increase in throughput.

At press time, IBM has developed integrated SSA chips, and a handful of hard disk vendors have equipment on the market; more penetration is expected over time. Early entrants include IBM, EMC, Silicon Systems, AT&T, Micropolis, and Connor Peripherals [12].

I/O and LAN Technology

The Fibre[22] Channel architecture combines the best aspects of *networks* and *I/O channels* into a single fiber-based device interface (coax and twisted-pair are also supported). FC is a multitopology, physical transport channel intended to serve as a common physical layer to carry multiple upper-layer protocols. FC supports SCSI-3, High Performance Parallel Interface (HIPPI), ATM, and IP. It operates at 100 MBps per link, making it the fastest I/O system at this time. (Speeds of 12.5, 25, and 50 MBps are also supported.) FC is an ideal match for the high-performance PCI local bus. Proponents see the ability of FC to support both I/O and network protocols simultaneously as the catalyst for winning favor among integrators. A single disk drive is expected to be capable of saturating a 20-MBps fast-wide SCSI interface by 1997, giving FC a window of opportunity at the channel level. As a network technology, however, doubts exist: At this time, it would appear that ATM will win the enterprise LAN, campus backbone LAN, MAN, and WAN races.

Fibre Channel provides a very high performance, multiprotocol interconnect that scales in cost and performance from supercomputing to individual disk drives. Supercomputers and high-end graphics processors still use HIPPI to access data. Mainframes use IBM's Enterprise Systems Connection (ESCON) with high performance HIPPI switches and ESCON Directors (ESCON switches) to achieve high-performance switched connectivity. Workstations and PC servers are limited by SCSI channels, with no comparable switches to route stored data to multiple systems. Fibre Channel solves these problems in an integrated manner, since HIPPI, ESCON, and SCSI commands and data are mapped onto FC and can be simultaneously transported on a single FC network [11]. In addition to handling the large data transfers supported by HIPPI, ESCON, and SCSI, networking protocols such as IP and ATM are being mapped onto FC. Proponents support the position that FC will provide network administrators with the ability to scale connectivity and performance with a relatively low entry cost approach. Figure 1.19 depicts FC's protocol stack.

---

[22] *Fibre* is the correct spelling of this standard since the supported media includes more than just fiber.

FC supports four data rates, three kinds of media, four transmitter types, three distance categories, three classes of services, and three possible fabrics; some view this as adding quite a level of complexity [21].

Standardization and interoperability profile work is being done by the ANSI X3T10 SCSI Technical Committee and the ANSI X3T11 Fibre Channel Technical Committee, as well as by industry ad hoc and promotional groups such as the HP-IBM-Sun Fibre Channel Systems Initiative group (FCSI), the Fibre Channel Association (FCA), and the Native Attach Drive Ad Hoc Working Group [11].

FC's I/O operations are based on the exchange of error-protected frames between nodes (using cyclic redundancy checking methods). As is the case in SSA, FC operates in full-duplex mode and requires no switches or terminators. FC supports several device interconnect topologies, including switched, point-to-point, and arbitrated loop (called *FC-AL*). FC-AL distributes switch logic to the individual devices and is a low-cost method of I/O attachment, particularly for disk drives. Devices can be removed from the connecting I/O network by using a bypass circuit. The application of a central hub for FC-AL provides a wiring framework that

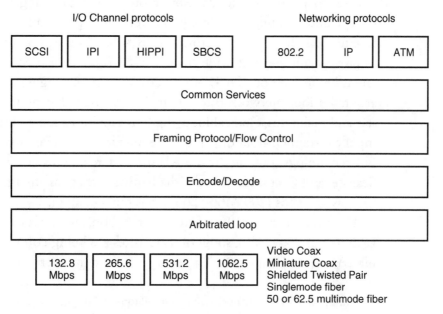

Figure 1.19 *The FC protocol stack.*

provides loop resiliency and network-friendly connectivity at 1 Gbps and a low-cost-bandwidth entry point; it also provides scalable, enterprisewide connectivity through the addition of FC switches (or *fabrics*) without obsoleting any existing equipment or wiring [11].

Fibre Channel is basically a point-to-point communication protocol: Connection or connectionless links are supported either through a switch or over the arbitrated loop. FC–AL is the simpler and least expensive of the two. In FC–AL no central switch is used to establish communication; instead, an arbitration protocol is established for each node to set up a communication channel. Once a node gains the right to communicate, it opens a connection to another node and full–duplex data transfer between the two participating nodes can start (up to 1.0625 Gbps). Other nodes in the loop act as repeaters. FC–AL supports up to 127 physical addresses.[23]

In addition to the high throughput, other advantages include distances from 30 m (on coax) to 10,000 m (on fiber). High levels of connectivity are supported via dual-ported full-duplex operation, which also supports fault tolerance. However, these advantages come at the expense of increased complexity, impacting software, firmware, and hardware.

SCSI mapped over 1.0625-Gbps Fibre Channel is being adopted by the storage and computing industry as a very high performance, serial interface for disk drives and for storage subsystems (this is more than a fivefold improvement over fast-wide SCSI). FC–AL is being used on disk drives and is being used on RAID and JBOD subsystems. Figure 1.20 depicts an arbitrated loop network through an FC–AL hub that provides a physical star and maintains a logical loop; the hub provides a means for reconfiguring the system without disrupting the operation of the network [11]. Early FC technology entrants included Sun Microsystems, Hewlett-Packard, Seagate, and Quantum. A lot of the hardware that has emerged operates at the lower speeds (200 Mbps; 266 Mbps for signaling).

The value of FC in the context of this book is that it can be used as a LAN technology. Also, some of its features are being considered for gigabit LANs (see Chapter 2). This topic is treated at length in Chapter 7.

---

[23] In general, Fibre Channel supports 6 million total physical addresses.

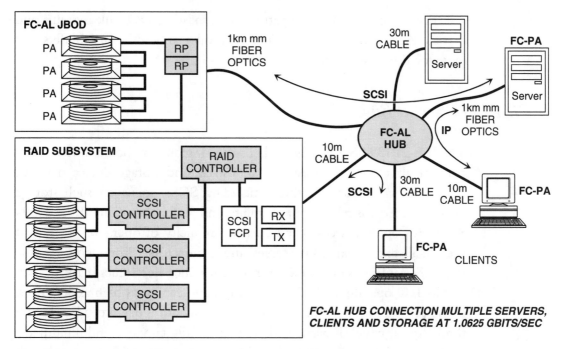

Figure 1.20 *Example of an FC-AL network configuration.*

PCMCIA

PCMCIA was originally defined as a memory card standard for portable computers; revisions of the standard added support for I/O cards. PCMCIA Flash memory cards are a nonvolatile mass storage media targeted for use in portable (and, more recently, desktop) computing environments. PCMCIA defines three form factors: type I (3.3-mm thick), type II (5-mm thick) and type III (10-mm thick). PCMCIA defines both the hardware specifications and the software layers required for compatibility [22].

A PCMCIA card contains an attribute memory that defines the functionality of the card, a gate array to control the internal operation of the card, and the Flash memory devices. The attribute memory contains the card information structure (CIS), which describes the card manufacturer, card density, and other important information. Another type of Flash memory storage card is the PCMCIA ATA card. These cards emulate IDE disk drives, so they are compatible with existing DOS file systems and drivers. ATA cards contain a microcontroller and additional buffers and logic for IDE disk emulation.

IEEE P1394 Serial Bus

The IEEE P1394 Serial Bus is a high-speed, global outboard device interconnect aimed at the consumer, industrial, and storage device markets. While other I/O architectures, including SCSI, focus on such storage device concerns as reliability, availability, and performance, the IEEE P1394 Serial Bus optimizes cost per port and simplicity of operation. Hence, it uses small and inexpensive cables and connectors. Proponents see the P1394 Serial Bus as being ideal for the consumer electronics market.

The I/O operates at 12.5 to 25 MBps, and can go as high as 50 MBps. It also provides support for isochronous bandwidth (guaranteed bandwidth throughput), which is particularly relevant to digital video and multimedia applications. The IEEE P1394 Serial Bus protocol maps SCSI-3 commands onto serial bus frames. IEEE P1394 Serial Bus includes protocols for bus arbitration, enabling it to support an evolution path from parallel-based SCSI I/O. In addition, the bus is dynamically reconfigurable, including the support of hot-plugging. The limitations of this I/O technology relate to the distance spanned (4.5 m) and the fact that the entire I/O bus is rendered busy when a frame is transmitted.

Early entrants in IEEE P1394 Serial Bus technology include Apple Computers, Sony, and Texas Instruments.

Other Technologies

A discussion of two technologies relevant to FC follows.

HIPPI

HIPPI is an established standard for system-to-system and system-to-peripheral environments. HIPPI was conceived at Los Alamos National Laboratory in the late 1980s, and was developed as a supercomputer technology. Standardization work by ANSI started in 1987, with standards published in 1991 (HIPPI is also an International Organization for Standardization standard). HIPPI has been widely adopted by the research community. HIPPI can support connections at 100 MBps or 200 MBps. HIPPI has built-in features for high-bandwidth network switching [23]. It can define multiple point-to-point channels between CPUs and from CPUs to storage systems. Table 1.8 provides a snapshot of some of the key HIPPI features. Some see a limited future for native HIPPI since it can be carried or supplanted by FC.

HIPPI over copper now allows several switches to be cascaded via local connections that can reach 200 m; multimode connections reach 300 m and single-mode connections (serial HIPPI) reach 10,000 m. If the switches are not equipped with a serial interface, then HIPPI fiber extenders are utilized. HIPPI is being used in data center networking, for internetworks (high-speed routers), for high-speed access to storage devices, and in workstation clusters. It is used for PC network adapters for the PCI bus; it is also used in conjunction with HIPPI switching fabrics.

Both the HIPPI Networking Forum and the ANSI X3T11 Technical Committee are working on activities to simplify the deployment of HIPPI—specifically, the development of a HIPPI management information base (MIB). Issues of practical network management (e.g., access to MIB via SNMP) as well as integration with commercial network management platforms, such as HP OpenView and Netview, are receiving attention from the vendor community. Capacity planning is also important. Interworking with LAN and WAN technologies, including ATM, SONET, and FC, was underway at press time. Transport of HIPPI protocol data units (PDUs) over an ATM network is via the use of AAL 5, as is the case for the support of other key data services[24] (the HIPPI PDUs are tunneled over the ATM network). For comparison purposes, it is worth noting that ATM switches now typically operate in the 2, 4, 8, 16, 20, 40, 80, and 320 Gbps ranges; HIPPI switches currently operate in the 12.8 to 25.6 Gbps range. Figure 1.21 depicts the HIPPI protocol stack [23].

There is also work on supporting HIPPI-over-FC and FC-over-HIPPI interworking. An ANSI standard has already been defined on how to send

---

[24] The reader interested in ATM may consult, for example, Reference [1].

Table 1.8 Key Features of HIPPI

Application	High-speed channel to connect peripherals such as disk and tape controllers. (IPI-3 is employed for host-to-peripheral communication, supporting high I/O throughput with minimal CPU overhead.)
Cabling	50-pair STP, single-mode and multimode fiber.
Circuit switching	Connection-oriented paths can be established through nonblocking switches. (Aggregate bandwidth = $n \times 1.6$ Gbps, where $n \geq 8$.)
Connection time	<1 μs for dedicated connection.
Cost per device adapter card	$2,000–$4,000.
Cost per switched port	$2,000 (down from $10,000 in early 1990s).
Distance	50 m point-to-point. Cascaded switches can be extended 200 m over copper, 300 m over multimode fiber, and 10,000 m over single-mode fiber (using Serial HIPPI—Serial HIPPI on multimode fiber supports 300 m).
Flow control (physical layer)	Credit-based system supporting reliable communication between devices operating at different speeds. (Source monitors *ready* signals and sends information only when destination is able to receive it.)
Implementations	Parallel HIPPI and serial HIPPI. (Serial HIPPI operates at 830 nm for multimode fiber and 1310 nm for single-mode fiber.)
Latency	160 ns.
Products	Switches, routers, mainframe and supercomputer channels, mass storage devices, and adapters for PCI workstations and PCs.
Protocol independence	Can handle "raw-HIPPI" (data formatted using HIPPI framing, without upper layers), TCP/IP protocol data units, and IPI-3 (Intelligent Peripheral Interface-3—used to connect peripherals to computers).
Signaling	Simple signaling sequences to set up connections: *request* (used by source to request connection); *connect* (used by destination to indicate that the connection has been established); and *ready* (used by the destination to indicate that it is ready to accept a stream of packets).
Speed	100 MBps (simplex) and 200 MBps (duplex on two links)—12 MBps sustained.
Standards	ANSI X3.183-1991 (physical layer [HIPPI-PH]). ANSI X3.222-1993 (switch control [HIPPI-SC]). ANSI X3.218-1993 (link encapsulation [HIPPI-LE]). ANSI X3.210-1992 (framing protocol [HPPI-FP]). ANSI/ISO 9318-3 (disk connections). ANSI/ISO 9318-4 (tape connections). ATM mapping work under way.
User applications	Supercomputers—digital signal processing of radar data, scientific visualization, storage, geological modeling, seismic analysis, molecular modeling and simulation, fluid analysis, structural analysis, animation, medical imaging, genetic mapping, and special effects for movies [21].

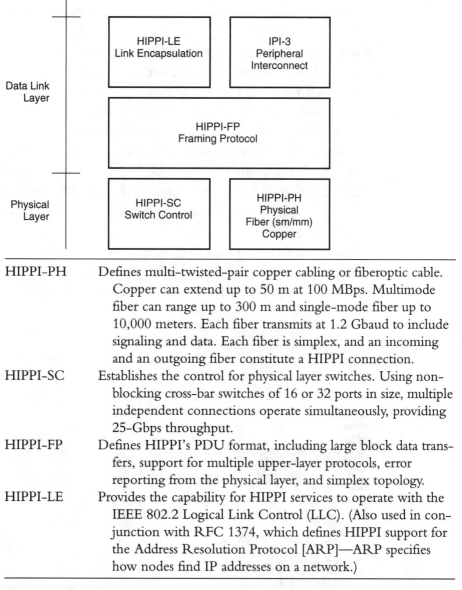

HIPPI-PH	Defines multi–twisted–pair copper cabling or fiberoptic cable. Copper can extend up to 50 m at 100 MBps. Multimode fiber can range up to 300 m and single-mode fiber up to 10,000 meters. Each fiber transmits at 1.2 Gbaud to include signaling and data. Each fiber is simplex, and an incoming and an outgoing fiber constitute a HIPPI connection.
HIPPI-SC	Establishes the control for physical layer switches. Using non-blocking cross-bar switches of 16 or 32 ports in size, multiple independent connections operate simultaneously, providing 25-Gbps throughput.
HIPPI-FP	Defines HIPPI's PDU format, including large block data transfers, support for multiple upper-layer protocols, error reporting from the physical layer, and simplex topology.
HIPPI-LE	Provides the capability for HIPPI services to operate with the IEEE 802.2 Logical Link Control (LLC). (Also used in conjunction with RFC 1374, which defines HIPPI support for the Address Resolution Protocol [ARP]—ARP specifies how nodes find IP addresses on a network.)

Figure 1.21 *HIPPI protocol stack.*

upper-layer FC PDUs over lower-layer HIPPI media; the other standards (how to transmit HIPPI PDUs over FC lower-layer media) is in process at press time.

At the simplest level, a HIPPI network consists of two computers with HIPPI channels linked by two 50-pair copper cables, supporting full-duplex (on two links) 100-MBps throughput at up to 25 m. As noted, cascading of HIPPI switches is allowed, as seen in Figure 1.22. HIPPI can also be used for LAN interconnection (RFC 1347 specifies how HIPPI can be used in the IP context). Many vendors of HIPPI equipment also have software drivers for TCP/IP. HIPPI can also be utilized in a hierarchical internetworking arrangement. For example, in a three-tier arrangement, Ethernet segments could be connected over FDDI backbones, which could in turn be connected over a HIPPI "network." Naturally, this is only one design alternative—another alternative would be to use ATM; see Figure 1.23.

HIPPI addresses are 24 bits long. Some switches work with 3-bit addresses; others with 4-bit. If 8-by-8 switches are deployed (which use the 3-bit addressing mechanism to identify one of the 8 inbound ports and

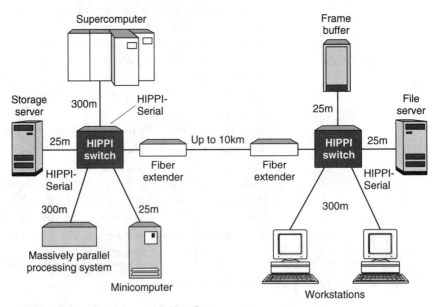

HIPPI = High performance parallel interface

Figure 1.22 Serial HIPPI implementation (via extenders).

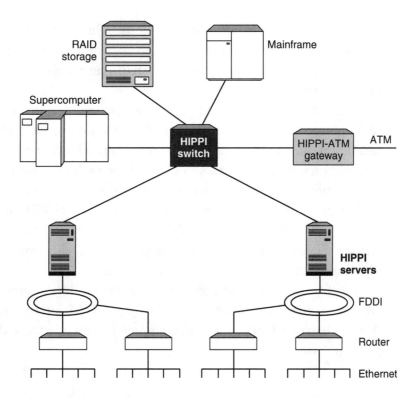

RAID storage

Mainframe

Supercomputer

HIPPI switch

HIPPI-ATM gateway

ATM

HIPPI servers

FDDI

Router

Ethernet

HIPPI = High performance parallel interface
RAID = Redundant array of expensive disks

Figure 1.23 *Use of a HIPPI switch.*

8 outbound ports), up to 8 switches can be cascaded (using up the 24-bit address space). Proponents also claim that HIPPI supports the lowest cost of switching per Mbps [23]—a $50,000 25-Gbps HIPPI switch has a figure of merit of $4/Mbps.[25]

Some in the industry have questioned whether HIPPI, which is a circuit-switched connection-oriented technology,[26] can handle datagram traffic efficiently—especially given the small size of TCP/IP PDUs [21]. Proponents argue that HIPPI's streamlined signaling sequences allow connections to be set up and torn down in less than 1 μs. Thus, a single port on a HIPPI switch can deliver hundreds of thousands of IP packets per

---

[25] A $20,000 1.6-Gbps Ethernet switch has a ratio of 12.5; a $120,000 3.2-Gbps FDDI switch has a ratio of 37.5; a $40,000 2.4-Gbps ATM switch has a ratio of 16.6.

second, outperforming IP routers or hosts, which typically would be the source of this traffic. In addition, switch latency is on the order of 160 ns; this puts the bottleneck on the end system's protocols rather than on the HIPPI switch.

Workstation clusters could replace supercomputers in the future, for such applications as animation, scientific visualization, and high-resolution imaging. Utilizing clusters, the researcher can achieve parallel processing, wherein each processor attacks a piece of the problem. But to achieve the desired performance, the interconnecting network must support the throughput that these arrangements demand. Here HIPPI can be used to: (1) link workstations and other hosts, (2) connect workstation to storage systems, (3) attach display devices supporting real-time visualization, and (4) interconnect the cluster to another network.

HIPPI products are available from over 60 vendors at press time. Early entrants in the new features of HIPPI (e.g., ATM interworking) include Avaika Networks Corp., Essential Communications, Nestar, and GTE Government Systems.

PCI

PCI is a high performance CPU bus. Driven by the requirements of computer-intensive applications, higher server performance, and increasing network bandwidth, the PCI remedies the shortcomings of other buses in the existing computer environment.[27] Initiated by INTEL and now supported by over 300 vendors, PCI was developed to fill two roles. On the desktop and at the server, it is a processor-independent backplane between high-speed peripherals, memory, and CPU. In networks, it is used to accommodate the traffic between interface cards and memory. PCI is defined for both 32- and 64-bit-wide operation. At 32-bits wide, it yields a peak bandwidth of 132 MBps, while running at 33 MHz. A defined 64-bit extension doubles the bus peak bandwidth to 264 MBps. Configuration transparency is provided by a connector standard that accepts both the 32-bit and 64-bit cards, and a defined autoconfiguration capability that lets

---

[26] ATM is also a connection-oriented technology.
[27] This discussion is based on Reference [23].

users install PCI boards without having to manually configure jumpers, DIP switches, or interrupts. PCI is interoperable with ISA, EISA, MCA, SBUS, and VME by having defined a standard bridging technique between these existing buses. PCI is also independent from processor technology, so that CICC and RISC processors all work with PCI.

PCI's bandwidth of 123-MBps throughput, together with its processor independence, extensibility, and interoperability with other buses make it the preferred bus architecture for gigabit network interface cards for PCs, servers, and personal workstations. In turn, this drives the throughput requirements for processor-to-mass-memory channels, as discussed earlier in the chapter.

I/O for the Rest of the Decade

As alluded to throughout this section, all of these I/O interfaces will likely be around for the rest of the decade. The new interfaces offer obvious advantages, but there is inertia for the existing technology due to the large installed base and the amortized software and hardware.

Fast-20 will not be in a position to displace ATA/IDE in the disk drive market as a whole, although it offers relatively high performance with minimal transition complexity. However, an increasing fraction of high-performance systems are employing it now. In spite of the fact that SSA and FC offer clear advantages in terms of performance and reduced bulk, they require nontrivial changes in hardware and software, all with cost implications. Developers of high-end storage-intensive systems will be able to justify the changes to these technologies, while others may not. As a consequence, deployment of SSA and FC will be small compared to Fast-20 in through 1998, but will very likely accelerate thereafter. Specifically, these interfaces will not migrate down to the PC market in this time frame, but will be confined to high-end workstation, server, and multimedia system environments. Observers hypothesize that SSA will be a viable commercial contender after 1998 only if it matches the performance of FC. PC manufacturers will likely drive the agenda since the upgrade cost for support of new I/O technology for disk drives is comparatively high. Therefore, the low cost, compatibility, and performance improvements offered by ATA-3 and Fast-20 are likely to make these standards the dominant choice up to 1998 [12].

This parenthetical discussion of I/O channels is aimed at exposing the planner to the fact that broadband really starts at the I/O level (even when *broadband* simply means a faster router). This general topic is not treated further in this book. However, as noted, FC has applicability beyond just I/O. It is a possible LAN technology, although ATM definitely has seen more LAN-centered development than FC. Hence, Chapter 7 focuses exclusively on FC.

How This Book Can Help

With all the networking options becoming available at the physical layer (e.g., to use 100Base-T or 100VG-AnyLAN), at the data link layer (e.g., to use frame relay, ATM, or LAN), and at the networking layer (e.g., to use MPOA or IP switching), the corporate planner may be bewildered. This book attempts to provide an early view on these topics to stimulate discourse and to facilitate planning and transitioning of enterprise, intranet, and interenterprise networks in the next few years.

References

1. D. Minoli and M. Vitella. *Cell Relay Service and ATM for Corporate Environments.* New York: McGraw-Hill, 1994.

2. D. Minoli and G. Dobrowski. *Principles of Signaling For Cell Relay and Frame Relay.* Norwood, Mass.: Artech House, 1995.

3. D. Minoli and T. Golway. *Designing and Managing ATM Networks.* Greenwich, Conn.: Manning/Prentice Hall, 1997.

4. D. Minoli and A. Schmidt. *MPOA.* Greenwich, Conn.: Manning, 1997.

5. D. Minoli. *Telecommunications Technologies Handbook.* Norwood, Mass.: Artech House, 1991.

6. D. Minoli and A. Alles. *LAN, ATM, and LAN Emulation Technologies.* Norwood, Mass.: Artech House, 1997.

7. D. Minoli and A. Schmidt, *Client/Server Over ATM.* Greenwich, Conn.: Manning, 1997.

8. W. K. Dawson and R. W. Dobison. "Buses and Bus Standards." *Computer Standards and Interfaces* 6 (1987): 403–425.

9. D. Minoli. *Telecommunications Technology Handbook.* Norwood, Mass.: Artech House, 1991.

10. S. Jones. "Ultra-SCSI Is the Bridge to Future Serial Interfaces." *Computer Technology Review* (Winter/Spring 1995): 84–87.

11. H. Chin. "Fibre Channel Offers Another Road to High-Speed Networking and I/O." *Computer Technology Review* (Winter/Spring 1995): 45–48.

12. L. Kubo. "SCSI Wars: Competing Technologies Prepare to Mount Major Offensives." *Computer Technology Review* (Spring/Summer 1995): 83–86.

13. M. Ferelli. "Is SCSI RAID's Only Interface." *Computer Technology Review* (Winter/Spring 1995): 93–97.

14. J. Warford. "SCSI Gets Faster with a Compatible Extension." *Computer Technology Review* (Winter/Spring 1995): 96–98.

15. J. M. Monti. "Single-Chip SCSI Comes to the PCI Local Bus." *Computer Technology Review* (Summer/Fall 1994): 105–107.

16. J. Molina. "Definition of RAID Levels." *Computer Technology Review* (August 1995): 38.

17. J. Dedek. "Message System/Command System Interactions in SCSI-2." *Computer Technology Review* (Summer/Fall 1994): 100–103.

18. B. Rudy and P. Chan. "Latest Advances in IDE, ATA and ATAPI Push SCSI to Higher Levels." *Computer Technology Review* (Winter/Spring 1995): 92–94.

19. *Computer Technology Review* (Winter/Spring 1995): 83.

20. J. Piven. "Will Enhanced IDE or SCSI Rule the Desktop?" *Computer Technology Review* (May 1995): 1.

21. D. Tolmie and D. Flanagan. "HIPPI: It's Not Just for Supercomputers Anymore." *Data Communications* (May 1995).

22. J. Kawaguchi. "PCMCIA Memory Cards: No Flash in the Pan." *Computer Technology Review* (Winter/Spring 1995): 41–44.

23. Promotional literature. Essential Communications, Albuquerque, N. Mex.

2 *LAN and Ethernet Switching*

Switched Ethernet at 10 or 100 Mbps has already become widely deployed in enterprise networks in the United States to support increased bandwidth to the client and server. Switched Token Ring technology has also seen some penetration. These switching technologies dedicate a LAN segment to a single user.[1] In so doing, they support not only increased bandwidth, but also lower interframe delay and delay variation, thereby making them more amenable to video and multimedia applications than shared-media LANs. Two other significant factors about this technology are that it utilizes embedded wiring, and the Network Interface Cards and the hub ports are very cost-effective. This technology will be very important well into the next decade. Therefore, this chapter provides a fairly inclusive treatment of the topic.

This chapter has two major modules. The first module,[2] consisting of "A LAN Technology Primer" and "Physical Devices Supporting Traditional LAN Interconnection," provides a review of 10- and 100-Mbps technology. It also covers bridging, which is the underlying technology for Layer 2 switching (the devices used to support switching are called *Layer 2 switches* [L2Ss]. This module is included to provide a self-contained primer on key LAN technologies. The second module, "LAN Switching," describes evolving LAN choices, particularly LAN switching, and the considerations that come into play in making appropriate deployment decisions.

---

[1] In effect, they follow the late 1980s approach utilized in Switched Multimegabit Data Service (SMDS), which entailed trivializing the IEEE 802.6 Dual Queue Distributed Bus (DQDB) protocol by putting a single user per access segment.
[2] This section is synthesized from References [1] and [2].

81

A LAN Technology Primer

As alluded to elsewhere in this text, after years of relative stability at the fundamental platform level, LAN technology is now seeing burgeoning activity in several radically new directions. A plethora of high-speed networking protocols has emerged in the past few years. These protocols have been proposed to accommodate evolving bandwidth-intensive multimedia and video applications. In addition to ATM LANs and emulated ATM LANs, technology now reaching the market includes switched 10-Mbps Ethernet, switched Token Ring, full-duplex Ethernet, 100-Mbps Ethernet (also called *fast Ethernet* [FE]), and switched 100-Mbps Ethernet, as well as other proposed systems (e.g., FC, gigabit LANs, IsoEnet/IEEE 802.9, FDDI II, and TP-PMD/CDDI/ANSI X3T9.5[3]). Corporate network planners are bewildered by the array of choices, as well as by the complexity of the new solutions, particularly in regards to systematic enterprisewide internetworking and network management. Ultimately, tens of billions of dollars are at stake within the North American corporate world, and millions of dollars are at stake within specific organizations.

The following material looks at 10Base-T (IEEE 802.3), 100Base-T (IEEE 802.3u), Demand Priority 100VG-AnyLAN (IEEE 802.12), and gigabit LANs as lower-layer LAN technologies. Extensions have been made over the years to IEEE 802.3 Ethernet since it was originally standardized in 1983. These extensions include Ethernet switching, full-duplex Ethernet, and, most recently, fast Ethernet (IEEE 802.3u.) In particular, Ethernet switching has been fairly successful at the commercial level.[4] Ethernet-based technologies (Ethernet switching, full-duplex Ethernet, and fast Ethernet, along with hubs, routers and adapters) tend to interwork with some degree of cohesion. Therefore, some see a continued, firmly rooted opportunity for this technology. Others see ATM as the

---

[3] This is FDDI over twisted-pair wire.

[4] Switching is a hubcentric form of Ethernet that takes advantage of the star topology now prevalent in LANs. It effectively eliminates Ethernet collisions by placing each workstation on its own segment (collisions occur only when the hub and the workstation both try to transmit information simultaneously).

next corporate LAN technology. Regardless of these arguments, it is worth noting that all the new protocols enumerated in this chapter support UTP wiring, thereby eliminating a potential first hurdle to deployment in the corporate landscape. However, there are significant technology, embedded-base, internetting, and management issues that impact the actual deployment prognosis.

Traditional Lower-Layer LAN Protocols

For LANs, Layer 1 and 2 functions of the OSIRM have been defined largely by the IEEE standards committees. The data link layer is split into two sublayers; these are the MAC, and the LLC (see Figure 2.1). The MAC specifies *access* to the physical transmission medium; this is done independently of the *physical characteristics* of the medium, but takes into account the topological aspects of the subnetwork. Different IEEE 802 MACs, including the ones in support of 100 Mbps, represent different protocols used for sharing the medium. The LLC sublayer provides a media-independent interface to higher-layer protocols. Upper layers use traditional standards such as TCP/IP and SPX/IPX, among others.

Carrier Sense Multiple Access/Collision Detect (CSMA/CD)

The Carrier Sense Multiple Access/Collision Detection (CSMA/CD) media access method for Ethernet is the means by which two or more stations share a common transmission medium. To transmit, a station waits

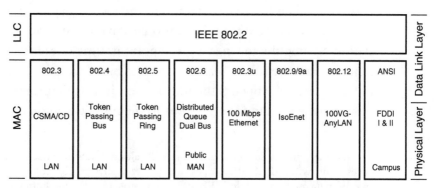

Figure 2.1 *Functions at specified LAN protocol levels.*

(defers) for a quiet period on the medium (that is, when no other station is transmitting) and then sends the intended PDU in bit-serial form. If, after initiating a transmission, the PDU collides with that of another station, then each transmitting station intentionally transmits for an additional predefined period to ensure propagation of the collision throughout the system. The station then remains silent for a random amount of time (backoff) before attempting to transmit again. This process is continued until the PDU is successfully transmitted.

Figure 2.2 depicts both the MAC Ethernet frame and the LLC frame that is encapsulated inside the MAC frame (in turn, the IP/IPX frame would be encapsulated inside the LLC frame).

Two key consequences of the CSMA/CD mechanism are *collision management channel overhead,* decreasing the maximum actual throughput well below the channel rate; and *unpredictable inter-PDU delay,* making it ill-suited for multimedia and video applications.

Token Ring

Token Ring technology is still widely used in IBM-based network environments. The IEEE 802.5 is nearly identical to (and is compatible with) IBM's Token Ring Network, although there are some minor differences, particularly at the physical level.[5] Token Ring technology operates at both 4 Mbps and 16 Mbps. The LAN uses token-passing methods, wherein a distinguished frame, called a *token,* is passed around the network (see Figure 2.3). Ownership of the token represents the right to transmit information. When a node receives a token and it has no information to send, it passes the token to the next station in the logical sequence. Each station in the network can hold the token for only a maximum amount of time. If the station owning the token at a given point in time does have information to send, it seizes the token, marks a specified bit of the token (making it a *start-of-frame sequence*), appends the information it has to transmit, and places the bits onto the medium for transport to the next station on the ring. There is

---

[5] IBM specifies a star configuration with all end stations attached to a multistation access unit, while IEEE 802.5 does not specify a physical topology. Also, IBM specifies twisted-pair medium, while IEEE 802.5 does not specify a medium type. There is also a difference in the RIF fields.

LLC Frame (LLC 1)

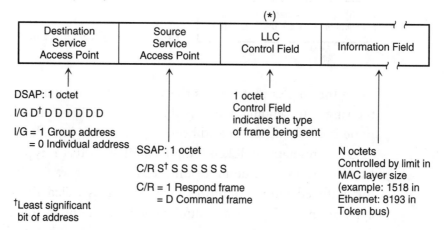

(∗) See Figure 2.5 and text for the SNAP extension format/mechanism.

MAC Frame

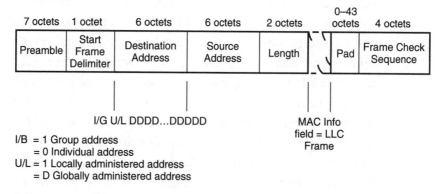

D bits comprise actual address (46 bits)

Figure 2.2 *LLC and MAC frames.*

no contention on a Token Ring network: While the information frame is traveling along the ring, there is no token on the network for other stations to use; hence, if these stations have information to send, they have to wait. In some cases, an *early token release* process is supported; here a new token can be released when frame transmission is completed.

The information travels along the ring until it reaches the destination. The destination copies the frame into its own internal buffers for local usage and processing. In the meantime, the frame continues to circulate

1 octet	1 octet	1 octet
Start Delimiter	Access Control	End Delimiter

Figure 2.3 *Token structure.*

along the ring until it reaches the sending station, which removes it from the ring. The sending station is able to scan the returning frame to determine if the frame has indeed been copied by the intended destination.

One advantage of Token Ring technology is that it is possible to calculate the maximum amount of time a station must wait before it can transmit buffered information; this environment is called *deterministic*. Video applications and process control benefit from this feature. In addition, Token Ring networks use a priority mechanism that enables priority-designated stations to have access to the network more frequently. The priority is supported via two PCI fields: the *priority field* and the *reservation field* (see Figure 2.4). With this mechanism, only stations with a priority equal to or exceeding the priority value shown in the token can seize the token and initiate transmission. Once a token is seized and transformed into an information frame, only LAN stations with a priority value higher than that of the transmitting station can issue a *reserve* indication for the token that will next circulate around the ring. When the next token is generated, it shows the higher priority that equates to the priority of the reserving station. Stations that increment a token's priority level are expected to decrement the token's priority to the previous level after they have concluded their transmission. As seen in Figure 2.4, there are two types of frames, besides the token. These are the *data frames,* which carry upper-layer PDUs, and the *command frames,* which carry only control information.

A selected station in a Token Ring network is designated to be the *active monitor* of the network (any station can assume this role). This station acts as a centralized source of timing information and supports a number of integrity maintenance functions. There is a need, for example, to remove a continuously circulating frame from the ring (this could happen when the sending station suddenly fails and is therefore unable to remove its own frame). Such a frame can preclude other stations from transmitting, thereby disabling the network. The monitoring station is responsible for detecting and removing such frames from the ring and launching a new

Start delimiter	Alerts each station of the arrival of a token, data, or command frame.★
Access control	Contains the priority, reservation, token bit, and monitor bit. The token bit differentiates the frame between token and data or command. The monitor bit is used to detect endlessly circulating frames.
Frame control	Indicates whether the frames contain data or control information. For control frames, this byte specifies the type of control and control information.
Address fields	IEEE-based 6-octet source and destination addresses.
Data field	Information field; length limited by the ring token-holding time.
FCS	Standard cyclic redundancy check field.
End delimiter	Indicates completion of the token and/or data or command frame.

★The first octet is distinguished from the rest of the frame by introducing a violation of the encoding scheme.

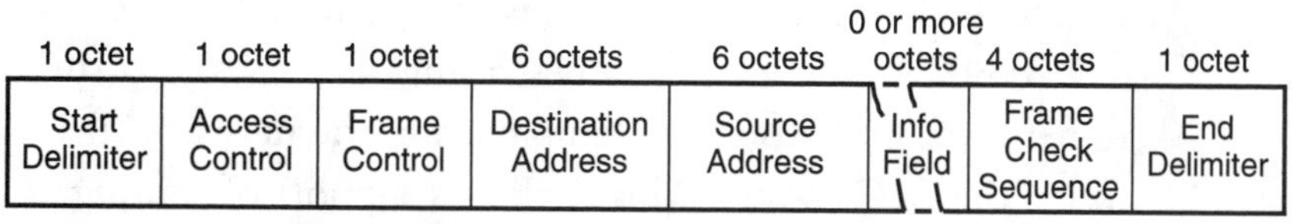

1 octet	1 octet	1 octet	6 octets	6 octets	0 or more octets	4 octets	1 octet
Start Delimiter	Access Control	Frame Control	Destination Address	Source Address	Info Field	Frame Check Sequence	End Delimiter

Figure 2.4 *Token Ring frame and description of fields.*

token. A process called *beaconing* detects and addresses some kinds of network faults. When a station detects a problem in the network, it issues a *beacon frame*. This frame defines a failure domain, which includes the station that issued the frame, its nearest upstream neighbor, and the facilities between these two stations. Beaconing starts the process of *autoreconfiguration,* wherein the nodes in the failure domain automatically initiate diagnostics and network reconfigurations.

Logical Link Control (LLC)

Service primitives are used to model the exchange of information between the different layers. A number of different service access points (SAPs) are defined in the IEEE 802 protocols; the SAP can be considered to be an addressable endpoint within a station that identifies a particular application or service.

The SAPs between the LLC layer and the higher layer are known as *LLC SAPs* (L-SAPs). Unacknowledged connectionless LLC 1[6] service supports only the higher-layer to LLC-sublayer L-SAP primitives necessary for *one instance of data transmission,* without prior connection establishment. The LLC Type 1 protocol frame has a 3-byte header comprised of an 8-bit destination service access point (DSAP) field, an 8-bit source service access point (SSAP) field, and an 8-bit control field, as shown in Figure 2.2; numbers for these fields are assigned by the IEEE (e.g., 00000000 = null SAP; 00100000 = SNA; etc.). After its standardization, however, an extension was made to the original IEEE 802.2 LLC Type 1 protocol to support the Sub-Network Access Protocol (SNAP).[7,8,9] It is an extension to the LSAP header just described. SNAP use is signaled by the value 170 in both the SSAP and DSAP fields in the LLC frame. Then the SNAP header follows; this header consists of 3 octets, showing an organization code, known as the organizationally unique identifier (OUI), followed by 2 octets showing a protocol ID (see Figure 2.5). This approach allows LLC encapsulation into 802.3, 802.4, and 802.5 frames, using the SNAP to specify the type of upper-layer protocol that is employed in the network (that is, encapsulated inside the LLC frame). This facilitates support of routing.

100Base-T LANs

The IEEE 802.3u[10] protocol maintains the 10-Mbps values of the IEEE 802.3 protocol and increases the throughput tenfold by speeding up the

---

[6] There are other types of LLC frames not discussed here.

[7] Some LANs continue to use the original Ethernet format. See Reference [3] for a discussion.

[8] RFC 1042, *A Standard for the Transmission of IP Datagrams over IEEE 802 Networks* (1988) indicated that from that point on, all IEEE 802.3, 802.4, and 802.5 networks should use the SNAP form of the LLC. In this implementation, DSAP and SSAP fields are set to 170 (indicating the use of SNAP) and SNAP is then assigned as follows: (1) 0 as organization code; (2) EtherType field—2048 for IP packets, 2054 for ARP packets, and 32821 for reverse ARP packets.

[9] Philosophically, one can interpret a set of header bits or bytes at a certain protocol level as another protocol sublayer; alternatively, these bits or bytes can be viewed as more protocol information that is part of the same protocol level.

[10] This discussion is based on IEEE 802.3u. It is, by design, limited in scope. Interested organizations should consult the original document for a much more inclusive discussion and for specification details.

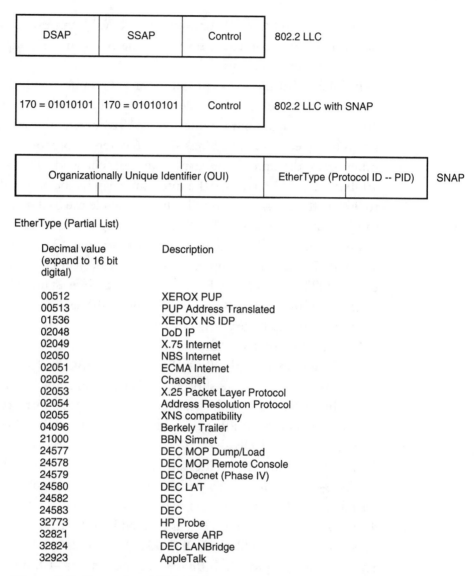

Figure 2.5 *Examples of SNAP extension.*

MAC [3]. 100Base-T couples the ISO/IEC 8802-3 CSMA/CD MAC with a family of 100-Mbps physical layers. While the MAC can be readily scaled to higher performance levels, new physical layer standards are required for 100-Mbps operation. 100Base-T uses the existing ISO/IEC 8802-3 MAC layer interface, connected through a media–independent interface (MII) layer to a physical layer device (PHY) sublayer such as

100Base-T4, 100Base-TX or 100Base-FX. With this extension, the IEEE 802.3 standard encompasses several media types and techniques for signal rates from 1 to 100 Mbps.

It follows that the MAC procedure, MAC frame length, MAC format, and maximum number of nodes in the LAN remain unchanged. As in 10Base-T, the maximum frame size is 1518 octets, the minimum frame size is 64 octets, and the address size is 6 octets. Protocol parameter values other than the interframe gap have not changed: The slot time for 100Base-T is 512 bit-times; the interframe gap has been scaled from 9.6 μs to 960 ns; the transmit attempt limit is 16, while the backoff limit is 10; and the jam size is 32 bits. IEEE 802.3u is designed to interwork with full-duplex techniques[11] and with Ethernet switching. One of the features of 100Base-T is that it can operate over the widely installed two-pair Category 3 UTP wiring. Fast Ethernet supports three standardized signaling systems—100Base-T4, 100Base-TX, and 100Base-FX. 100Base-TX supports two-pair Category 5 UTP. The IEEE published the finalized IEEE 802.3u fast Ethernet specification in late 1995.

100Base-T extends the ISO/IEC 8802-3 MAC to 100 Mbps. The bit rate is faster, bit times are shorter, packet transmission times are reduced, and cable delay budgets are smaller—all in proportion to the change in bandwidth. This means that the ratio of PDU duration to network propagation delay for 100Base-T is the same as for 10Base-T. As noted, the IEEE 802.3u standard specifies a family of physical layer implementations. 100Base-T4 uses four pairs of ISO/IEC 11801 Category 3, 4, or 5 balanced cable. 100Base-TX uses two pairs of Category 5 balanced cable or 150-Ω shielded balanced cable, as defined by ISO/IEC 18801. 100Base-F uses two multimode fibers. FDDI (ISO 9314 and ANSI X3T12) physical layers are used to provide 100Base-TX and 100Base-FX physical signaling channels, which are defined under 100Base-X. For comparison purposes, Table 2.1 depicts wiring schemes for a number of common LANs, including 100-Mbps LANs.

---

[11] In a full-duplex operation one no longer has flow or congestion control. Therefore, the impact of a PDU loss is more significant than a collision because it has to be resolved at a higher layer of the protocol stack (e.g., TCP). For this reason, some argue that while full-duplex operation has its place for switch-to-switch or server-to-hub connections over fiberoptic media, it is not appropriate for 100-Mbps connection to the desktop [4].

Table 2.1 Wiring Schemes for a Number of Common LANs

LAN Technology	WIRING PAIRS			WIRING TYPES			
	1 Pair	2 Pair	4 Pair	Category UTP	Category UTP	Category UTP	STP
IEEE 802.3							
Ethernet—10 Base-T		x		x		x	x
IEEE 802.12							
Demand priority—							
100VG-Any			x	x	x	x	x
IEEE 802.3u							
Fast Ethernet—							
100Base-TX		x				x	x
Fast Ethernet—							
100Base-T4			x	x	x	x	x
Fast Ethernet—							
100Base-FX	x						
Fast Ethernet—							
100Base-T2		x		x	x	x	x
IEEE 802.9							
IsoEnet		x		x		x	
ANSI X3T9.5							
TP-PMD—CDDI		x				x	x
ATM Forum							
25-Mbps ATM		x		x	x	x	x
ATM CAP-4		x		x	x	x	
ATM CAP-16		x		x	x	x	
ATM 155-Mbps							
Category 5		x				x	x
ATM 155-Mbps							
Category 3		x		x	x	x	

A concern about 100Base-T technology relates to LAN topology considerations and the possible need to rewire a portion of the enterprise. 10Base-T supports a diameter (the farthest node–to–farthest node distance) of 2500 m. By contrast, in a single segment, 100Base-T has a maximum collision domain, including repeaters, of only 205 m (half-duplex operation, balanced copper links, and no margin). Note that a 14-ft slab-to-slab floor equates to about 4 m, so in theory a 50-story backbone could be supported; however, there is typically quite a bit of lateral-run distance to be covered. The (reduced) diameter is a function of the cable type, the signaling system used, and the repeater configuration along the path. The distance specified in IEEE 802.3u takes into consideration: (1) the time needed for the MAC procedure to detect a round-trip collision using the CSMA/CD mechanism, (2) the round-trip delay between two nodes when the signal propagates through two repeaters using Category 5 UTP, and (3) the time it takes a station to detect a collision from a station on the farthest end of the segment—this time must be within 512 bit-time slots (64 bytes equating to the minimum Ethernet frame size), since the sending station is required to have not finished transmitting the entire frame before a collision fragment is detected. Switching and full-duplex techniques can be employed to extend the distance. Ethernet switches partition cable runs into separate collision domains: The switch regenerates each PDU before sending it on to another 10- or 100-Mbps network. Specifically, PDUs are forwarded only to their destination address, instead of being broadcasted to all segments that are attached to the switch. Full-duplex eliminates the collisions mechanism, thereby also extending the reach.

Another concern about 100Base-T technology relates to the fact that the specification sets a two-hub-per-repeater maximum (with an interrepeater distance not to exceed 5 m), for latency reasons. Repeater sets are an integral part of any 100Base-T network with more than two end systems in a collision domain. They extend the physical system topology by coupling two or more segments. Repeaters are permitted within a single collision domain to provide the maximum path length. The two-repeater limit implies a single-level hub structure, and since fast Ethernet hubs now on the market support up to 24 ports per hub, this results in a network with at most 48 stations. However, new repeater chips under development will facilitate the introduction of hubs supporting 100 fast Ethernet ports per hub; these hubs can then be connected to form a two-hub network of approximately 200 nodes. It is also expected that companies will introduce stackable hubs where the cumulative latency for the stacked multiple hubs will be no greater than the two-repeater hub latency limit. A higher number of nodes can be accommodated if switches are used in conjunction with the hubs or if a mixed-media repeater (for example, 100Base-TX and 100Base-FX fiber) is used.

Backward Compatibility

Fast Ethernet has an option called *autonegotiation* (AUTONEG) to support backward compatibility to 10Base-T. This procedure allows two NICs to choose the highest common performance capability. AUTONEG provides a linked device with the capability to (1) detect the abilities (modes of operation) supported by the device at the other end of the link, (2) determine common abilities, and (3) configure for joint operation. AUTONEG is performed out of band using a pulse code sequence that is compatible with the 10Base-T link integrity test sequence.

Fast Ethernet equipment has the ability to support both 10- and 100-Mbps speeds, in addition to a choice of half- and full-duplex operations. However, the embedded base of Ethernet equipment obviously supports only 10 Mbps. Therefore, this kind of autonegotiation is important. The downside is that some users will be able to run their applications at 100 Mbps, while others will be restricted to 10 Mbps. Each fast Ethernet NIC sends fast-link pulses (FLPs) to the other NICs through the hub during power-up (or on command) to accomplish this handshaking. After the highest common performance capability is selected, FLPs cease to be transmitted. The FLPs contain encoded information[12] that is used to control the AUTONEG function; the FLP burst lasts up to 2 ms and is encapsulated in the place of the single 10Base-T link integrity test pulse. The

---

[12] FLPs are separated by 16 ± 8 ms and contain 33 data-pulse positions. The clock represents 17 pulses and the other 16 represent capability information (d0d1d2d3d4d5d6d7d8d9d10d11d12d13d14d15). d0 to d4 represent the *Selector Field;* d5 to d12 represent the *Technology Ability Field;* d13 is the *Remote Fault* (RF) bit; d14 is the *Acknowledge* bit (used to indicate that a device has successfully received its link partner's *Link Code Word*); and d15 indicates a *Next Page* presence.

There are 32 combinations in the s4s3s2s1s0 selector field. The values are: 00000 = reserved for future autonegotiation development; 00001 = IEEE 802.3; 00010 = IEEE 802.9; 11111 = reserved for future autonegotiation development.

There are 256 combinations in the 8-bit a0a1a2a3a4a5a6a7 Technology Ability Field; however, only a small subset is used. This field contains information about the supported technologies, since two NICs might have multiple abilities in common. The codepoints are: 10000000 = 10Base-T; 01000000 = 10Base-T full-duplex; 00100000 = 100Base-TX; 00010000 = 100Base-TX full-duplex; 00001000 = 100Base-T4. 000000100, 000000010, and 00000001 are reserved for future technology. This information, however, is conveyed via a prioritization scheme. The priorities (listed from the highest to the lowest: 100Base-TX full-duplex, 100Base-T4, 100Base-TX, 10Base-T full-duplex, and 10Base-T) are different than the ordering just described for the technology ability field bit assignments.

FLP signals are compatible with 10Base-T normal link pulses (NLPs), but contain significantly more information encoded within the burst.[13]

Consider the handshake between a NIC that is autonegotiation-capable and a 10Base-T NIC. When the autonegotiation-capable NIC or terminal powers up it sends out FLPs. The 10Base-T terminal may transmit NLPs or actual traffic. In either case a 10Base-T mode is determined. Now consider the handshake between two NICs that are autonegotiation-capable. Upon powering up, both start transmitting FLPs. Upon receiving three consecutive and consistent FLPs from the far end, the capabilities of the far end are recognized by the local end. The *Acknowledge* bit[14] is set, and at least four more FLPs are transmitted. For a total handshake, the receiving station expects to receive at least two consecutive FLPs with the Acknowledge bit set. The highest level of common performance is, thereby, selected.

Media-Independent Interface (MII)

Three important compatibility interfaces are defined in 802.3u within what is architecturally the Physical Layer [3]:

1. *Medium-Dependent Interface (MDI)*. To communicate in a compatible manner, all stations need to adhere rigidly to the exact specification of physical media signals defined in 802.3u. LANs require complete compatibility at the physical medium interface (that is, the physical cable interface).

2. *Attachment Unit Interface (AUI)*. It is anticipated that most user devices will be located some distance from their connection to the physical cable. A small amount of circuitry then needs to exist in the *Medium Attachment Unit* (MAU) directly adjacent to the physical cable, while the majority of the hardware and all of the software will be placed within the user device (e.g., PC or server). The AUI is defined as a second compatibility interface. While conformance with this interface is not strictly necessary to ensure

---

[13] The encoded data of the FLP contains the *Basic Page*. Additional pages are supported using the *Next Page* exchange.
[14] Bit d14.

communication, it is recommended in the IEEE standard, since it allows maximum flexibility in intermixing MAUs and end systems. The AUI may be optional or not specified for some implementations that are expected to be connected directly to medium (and so do not use a separate MAU or its interconnecting AUI cable). The *Physical Layer Signaling* (PLS) and *Physical Medium Attachment* (PMA) are then part of a single unit, and no explicit AUI implementation is required.

3. *Media-Independent Interface (MII).* It is anticipated that some end systems will be connected to a remote PHY, and/or to different medium–dependent PHYs. The MII is defined as a third compatibility interface in IEEE 802.3u. While conformance with implementation of this interface is not strictly necessary to ensure communication, it is recommended in IEEE 802.3u, since it allows maximum flexibility in intermixing PHYs, and end systems (thus, the MII is optional).

The MII provides the interconnection between the MAC sublayer and the physical layer devices. It supports 10- and 100-Mbps datarates through independent nibblewide (4–bit) transmit and receive datapaths; transmission occurs one nibble at a time.[15] The MII is designed to mask differences among the various physical devices to the MAC.[16,17] The *Reconciliation Sublayer* provides a mapping between the signals provided at the MII and the MAC/PLS service definition. The MII interface has the following characteristics [3]:

1. Can support both 10- and 100-Mbps datarates.
2. Data and delimiters are synchronous to clock references.
3. Provides independent 4–bit-wide transmit and receive datapaths.

---

[15] Given bits b0b1b2b3b4b5b6b7, where b0 is the least significant bit (sent first) and b7 is the most significant bit (sent last), then b0b1b2b3 is sent as the first nibble with n0n1n2n3 as the word; b4b5b6b7 are sent as the second nibble with n0n1n2n3 as the word.

[16] This is functionally similar to the UTOPIA interface in ATM.

[17] MII also provides interconnection between PHY layer devices and station management (STA).

4. Uses TTL signal levels, compatible with common digital CMOS ASIC processes.

5. Provides a simple management interface.

6. Can drive a limited length of shielded cable.

100Base-T Options

100Base-TX uses the physical layer definition from copper-based FDDI over two-pair Category 5 UTP. One pair is used to transmit and the other to receive (it can run either half-duplex—receive or transmit—over two pairs, or full-duplex—receive and transmit—over four pairs). Since it uses a clock rate of 125 MHz (along with 4B/5B encoding, for an effective bandwidth of 100 Mbps), it needs careful design to satisfy the FCC's Class B emission standards. In cable plans based on Category 3, this system requires rewiring.

100Base-FX is the same as 100Base-TX except that it runs over a fiberoptic medium, eliminating emission problems.

100Base-T4 uses four-pair Category 3, 4, or 5 UTP. It uses a ternary encoding scheme, 8B/6B, where 8 bits are encoded in 6 three-valued symbols (−1, 0, +1). It allocates those signals among three pairs of wire for information exchange and uses the fourth pair for collision detection. It has a signal rate of 25 Mbaud and operates in half-duplex mode; also, because of near-end cross talk (NEXT), it cannot be used in older cable plants.

100Base-TX and 100Base-FX have seen rapid time-to-market because of the reuse of FDDI components.

100Base-T2

An IEEE 802.3y task force was formed in 1995 to develop a 100Base-T2 transceiver specification capable of operating at 100 Mbps over two-pair Category 3, 4, and 5 UTP wiring at full-duplex and over shielded twisted-pair (STP). A number of different technology proposals are under evaluation. The proposals are based on Ternary Partial Response (TPR) and Carrierless Quadrature Amplitude Modulation (C-QAM). All of these proposals use Digital Signal Processing (DSP) principles, such as bandwidth-efficient multilevel coding, channel equalization, spectrum shaping, echo/NEXT cancellation, and adaptive signal processing techniques. For a given bit rate, multilevel coding decreases the bandwidth requirement for the transmission channel. Conversely, given certain channel characteristics, multilevel coding maximizes the achievable datarate. The use of these technologies for high-speed data transmission over UTP is a result of its attenuation, roughness, and temperature/crosstalk coupling characteristics over Category 3 wire, as well as the need to keep down radiated emission levels to conform to FCC regulations [5].

Commercial Venues

The interests of FE have been advanced by the Fast Ethernet Alliance, headquartered in San Francisco, California.

100VG-AnyLAN

The IEEE 802.12[18] specification defines the protocol and compatible interconnection of data communication equipment via a repeater-controlled, star-topology LAN using the Demand Priority access method. 100VG-AnyLAN technology can be used over four pairs of Category 3, 4, or 5 wire. It employs the Ethernet framing format, but replaces CSMA/CD with a MAC method based on prioritized bandwidth demand (reservation). 100VG-AnyLAN takes the view that, given the hubcentric environment in LANs (with a move away from the original distributed bus approach), the hub can be further endowed with bandwidth-management features. With Demand Priority, the NIC notifies the hub that it has a transmission pending. The hub, in turn, authorizes transmission if it is idle.

The purpose of the IEEE 802.12 protocol is to provide a higher-speed LAN with deterministic access, priority, and optional filtering. The design goals of the 100VG-AnyLAN protocol are as follows [6]:

---

[18] This discussion is based on IEEE 802.12. It is, by design, limited in scope. Interested organizations should consult the original document for a much more inclusive discussion and for specification details.

1. Provide a minimum datarate of 100 Mbps.

2. Provide smooth migration from ISO/IEC 8802-3 and ISO/IEC 8802-5 LANs.

3. Support either ISO/IEC 8802 or ISO/IEC 8802-5 frame format and MAC service interface to the LLC.

4. Support a cascaded star topology over twisted-pair and fiberoptic generic building wiring.

5. Allow topologies of 2.5 km and greater with three levels of cascading.

6. Provide a physical layer bit error rate of less than 10^{-8}.

7. Provide fair access and bounded latency.

8. Provide two priority levels, normal and high priority.

9. Provide a low-latency service through high priority for support of multimedia applications over extended networks.

10. Support an option for filtering individually addressed PDUs at the repeater to enhance privacy.

11. Support network management to monitor network performance, isolate faults, and control network configuration.

12. Enable low-cost implementation and high levels of integration.

13. Provide for robust operation by testing the physical layer connection before allowing an end node to enter the network and by removing disruptive nodes.

In theory, Demand Priority can sustain 96 percent channel utilization, while Ethernet saturates at 80 percent (and, in practice, the actual utilization is kept well below 40%). As noted, it also has a two-level prioritization scheme, where high-priority requests from the NICs are serviced before normal-priority requests; if queued, the high-priority traffic is processed before normal requests are again serviced [4]. This feature is useful for multimedia support.

100VG-AnyLAN has a complex PMD, using all four pairs, to support quartet channeling, data scrambling, and 5B/6B encoding. The *quartet channeling* allocates the data among the four pairs. *Data scrambling* and

5B/6B encoding reduce RF emissions and crosstalk. This technology can utilize older cable plants. It cannot support full-duplex over Category 3, 4, or 5 UTP, but it can with STP and fiberoptic cable.

LAN Topology

The Demand Priority medium access protocol provides a means by which stations (end system) can communicate with each other over a centrally controlled LAN that offers a choice of several different link media, including 100-Ω balanced cable (4-UTP and 2-UTP), 150-Ω shielded balanced cable (STP), and optical fiber.

The Demand Priority access method has been defined to allow considerable flexibility in the network topology. The Demand Priority LAN architecture utilizes three structural components: end nodes, repeaters, and network links.

- *End nodes* are typically PCs, departmental computers, user workstations, peripherals, or file servers. They might also be special devices, such as bridges to other LANs.
- *Repeaters* are the network controllers and are configured with two or more local ports to allow connection to end nodes or other repeaters. They can also be configured with an optional cascade port, which is reserved for connection to repeaters only.
- *Link segments* provide the interconnection medium between a repeater and its connected end nodes or other repeaters.

Larger topologies can contain several levels of repeaters interconnected in a cascade. Each repeater is typically connected to one or more end nodes, and can be connected to one or more repeaters. Lower-level repeaters and end nodes are connected to local ports. Higher-level repeaters must be connected to a cascade port. Interconnection between two repeaters using only local ports is not allowed. The topmost repeater in the cascade is designated the *Level 1* repeater. Repeaters in each succeeding lower level in the cascade are designated by the number of links between them and the root repeater by the following equation:

Repeater level = Number of link segments away from the root repeater + 1

(All repeaters on the same level are designated with the same level number.)

The connection of a Demand Priority network to another LAN is accomplished via an appropriate bridge: Each bridge is connected to a local port on the Demand Priority repeater, and is treated as if it were an end node on the network. PDUs can be sent to an individual end node, to groups of end nodes (*group addressing*), or to all end nodes on the network (*broadcast addressing*). Repeater ports connected to end nodes can be configured to support either of two addressing modes: *privacy mode,* where packets are sent to addressed end nodes only, or *promiscuous mode,* where the port transmits all traffic.

Basic Operation

Demand Priority access is a priority-based, round-robin arbitration method wherein the central network controller (the repeater) regularly polls its connected ports to determine which have transmission requests pending and whether each transmission request is normal-priority (e.g., for data files) or high-priority (e.g., for real-time voice, video, or data) [6].

Round-Robin Polling   Round-robin polling provides all end nodes with access to the network during each round-robin cycle. The repeater maintains two next-port pointers, one for normal-priority and one for high-priority requests, that keep track of the next ports to be serviced. All ports are polled at least once per PDU transmission to determine which have requests pending.

A high-priority request allows an end node to obtain permission to send sooner than its normal-priority cycle position. High-priority requests are serviced before normal-priority requests, but do not cause a normal-priority transmission in progress to be interrupted.

In the repeater, timers monitor the normal-priority requests at each port to ensure that abnormally large high-priority traffic levels do not preclude normal-priority traffic. Normal-priority traffic requests that have been pending for approximately 250 ms are automatically elevated to high-priority and are serviced in port order as indicated by the high-priority next-port pointer. Priority promotion is not serviced in port order as indicated by the high-priority next-port pointer.[19]

Data Transmission   PDU transmission follows a request/grant sequence wherein the request is initiated by the sending end node and the issuance of the grant is controlled by the repeater. When PDU transmission is occurring [6]:

1. The repeater polls all local ports to determine which end nodes or lower-level repeaters are requesting to send a PDU and whether the request is normal- or high-priority. If an end node has a PDU ready to send, it transmits either a *Request_Normal* or *Request_High* control signal. Otherwise, the end node transmits the *Idle_Up* control signal.

2. Once the repeater has finished transmitting the PDU, it selects the next end node with a request pending by sending *Grant* to the selected end node. *Idle_Down* is sent to all other end nodes. PDU transmission begins when the end node detects the *Grant*.

3. The repeater then alerts all end nodes on the network (other than the sender) of a possible incoming PDU (by transmitting an *Incoming* control signal) and decodes the destination address as it is being received.

4. When an end node receives the *Incoming* control signal, it prepares to receive a PDU.

5. After the repeater has decoded the destination address, the PDU is sent to the addressed end node(s) and to any promiscuous ports. Communication to nonaddressed, nonpromiscuous ports is inhibited during PDU transmission (the repeater sends *Idle_Down* to these ports during packet transmission).

---

[19] Priority promotion is not intended to occur during normal network operation. When an abnormal situation occurs—for example, an end node continuously making priority requests—priority promotion allows normal-priority requests to be serviced until network management locates and isolates the source of the problem.

Operations in Cascaded Networks   Transmission in cascaded networks follows a similar procedure to operations in a single-repeater network [6]. The interconnected repeaters act essentially the same as a single large repeater. All traffic is sent to each repeater, and each polls its active ports for requests at least once per PDU transmission. The Level 1 (root) repeater has primary control of the PDU sequencing. Lower-level repeaters with requests pending issue requests to the next higher-level repeater in the same manner as an end node.

Transmission Errors   Error detection mechanisms in the data code and the cyclic redundancy check appended to each PDU enable repeaters and end nodes to detect such errors. Invalid PDUs received by the repeater are marked by substituting an invalid packet marker (IPM) for the end-of-stream delimiter (ESD) in the outgoing PDU's end of frame sequence. Invalid PDUs received by the destination end node can be discarded. Recovery from transmission errors is the responsibility of upper layers in the protocol stack.

Protocol Model End Node Architecture

The four sublayers of the 802.12 reference model for the Demand Priority end node correspond to the two lower layers of the ISO 7498 model for Open System Interconnection [6]. The upper sublayer in the end node is typically an ISO/IEC 8802-2 Class I LLC supporting Type 1 unacknowledged, connectionless-mode transmission, or an ISO/IEC 8802-2 Class II LLC supporting Type 2, connection-mode transmission. The MAC sublayer is compatible with two MAC formats and interfaces: the MAC-frame format and MAC/LLC interface of the ISO/IEC 8802-3 standard, and the MAC-frame format and MAC/LLC interface of the ISO/IEC 8802-5.2 standard.

The functions intended for the ISO PHY are encompassed by two sublayers: the *Physical Medium Independent* (PMI) and the *Physical Medium Dependent* (PMD). The MII (previously defined in the discussion of IEEE 802.3u) is defined as a logical interface between the PMI and the PMD. An optional physical definition of the MII allows the interchange of PMDs supporting different physical links. The MDI is usually a fully specified physically exposed connection between the PMD and the link medium. However, in some cases, a MAC is embedded in another device without an exposed MDI; in this case, the physical layer functions are logically, but not physically, present.

Other Facilities

A summary of other features and functions of the protocol model follows.

MAC Functions   The MAC accepts transmit requests and provides receive indications for PDU traffic between the end node and other end nodes. The MAC controls frame format composition and decomposition. It checks for transmission errors in received frames and initiates controls to the PMI.

RMAC Functions   The RMAC is the primary control sublayer of the network repeater. The RMAC accepts transmit requests from the end node, arbitrates PDU transfer sequences, interprets destination addresses, and routes incoming PDUs to the proper outbound ports. It also supplies a subset of the functions (such as PDU error checking) provided by the MAC in an end node.

PMI Functions   The PMI supplies transmit control state indications to the PMD, provides data scrambling and encoding for outgoing frames, and adds preambles, as well as the start and end-of-stream sequences, to the data streams prior to transmission. In the receive mode, the PMI accepts receiver control state indications from the PMD, strips the preamble and the start and end-of-stream sequences, and provides data descrambling and decoding for incoming frames.

Medium-Independent Interface   As previously noted, the MII is a local medium-independent interface between the PMI and the PMD. PMD service primitives are passed across this interface. The optional physical implementation of the MII provides interchangeability of PMDs and allows easy reconfiguration of end nodes and repeater ports to support multiple link media.

PMD Functions   The PMD provides control signal generation and recognition, datastream signal conditioning, clock recovery, and channel multiplexing appropriate to the link medium.

Medium-Dependent Interface   The MDI is the physically exposed connection between the PMD and the link segment. The PMD and MDI must both support the same link medium. MDI functions are defined for 4-UTP, STP, and LED fiberoptic link media.

Commercial Venues

The interests of 100VG-AnyLAN have been advanced by the 100VG-AnyLAN Forum, headquartered in North Highlands, California.

Gigabit LANs

The computer industry is experiencing an explosion of new networking and interconnection strategies that operate at or near 1 Gbps. Committees on Super High-Performance Parallel Interface, IEEE 1394.1, IEEE 802.12, and IEEE 802.3 all have projects underway to define inteconnects at this speed, and there are other alternatives being developed outside of the ANSI and IEEE standards organizations, as well. This heightened level of activity is being driven by technical and economic feasibility, as well as by real market need.

The 802.3 Working Group of the IEEE 802.3 LAN/MAN Standards Committee (LMSC) has begun work on a project known as *gigabit Ethernet,* which is a 1000-Mbps extension of the 802.3 Carrier Sense Multiple Access with Collision Detection based on the Local Area Network standard.

The project, which is expected to be designated as P802.3z, will build on the working group's experience on scaling the 802.3 standard to higher speeds, as reflected in the publication of the 100Base-T standard (IEEE Std 802.3u-1995), which extends the operating speed of CSMA/CD networks to 100 Mbps. The new project will also make use of the Full Duplex and Flow Control extensions to 802.3 that are being drafted by the 802.3z Task Force.

The physical signaling protocol for the gigabit Ethernet will be derived from the ANSI/ASC X3 Fibre Channel FC-1 and FC-0 specifications. The 802.3 working group has set an objective that the datarate of this new standard should be 1000 Mbps, as measured at the interface between the data link layer and the physical layer. This will require some deviation from the parameters of the FC-0 specifications, since FC-0 defines a datarate of 850 Mbps measured at the equivalent interface.

More information about this project can be found at the IEEE Standard Web Site, at ftp://stdsbbs.ieee.org./pub/802_main/802.3/gigabit.

A high-speed supplement currently under development, IEEE Std 802.12-1995, will provide gigabit transfer rates for demand-priority shared-media LANs while maintaining the 10-Mbps campus topology limits (8-km network diameter with up to 5 levels of cascade repeaters). The new gigabit network will utilize the current Demand Priority MAC protocol, with timing parameters scaled to the new operating rates. The new physical layer will be based on the Fibre Channel 8B10B link protocol. Link media will be ISO/IEC 11801: 1995-compliant optical fiber, either single-mode or multimode, depending on link-length requirements. Half-duplex burst-mode transmission will be provided on shared media. Full-duplex transmission (currently in development under a separate effort) will be available for dedicated links.

Two transfer rates are under consideration for the fiberoptic physical layers: 1.063 Gbps and 1.25 Gbps. The final decision will consider customer link-length requirements, the bandwidth and distance limitations of the optical fiber (the higher transfer rate could reduce the maximum link length by approximately 18 percent), and whether 1.25-Gbps FC chips can be reasonably produced.

A half-gigabit transfer rate is also under consideration for four-pair Category 5 (or better) copper media. The proposed physical layer is based on a multilevel transmission coding scheme, wherein simultaneous control and data transfer is accomplished through use of a 3 + 1 (3 data plus 1 control) link configuration. Half-gigabit optical fiber transmission will also be defined.

A web site on the supplement is under development at pat@hprnd.rose .hp.com.

There is already keen commercial interest in the technology. The gigabit NICs are being targeted at $2500 per port. Early entrants included 3Com, Bay, Cabletron, Cisco, Digital Equipment Corporation, Hewlett-Packard, Intel, Ipsilon, Madge, Nbase, Packet Engines, Prominet, Sun, XLNT Designs, and Xylan. An industry forum, the Gigabit Ethernet Alliance, was formed in 1996 to advance this cause. Some put the market at $3 billion by the year 2000, although it is likely going to be much smaller. ATM continues to be a strong contender in this broadband market, with over 12 years of standards and product development.

Physical Devices Supporting Traditional LAN Interconnection

This section reviews traditional physical devices that have been employed to achieve LAN interconnection. Figure 2.6 depicts, at a general level, all of the relay equipment utilized for internetworking.

Repeaters are devices that amplify signals in order to increase the physical range of the LAN. A repeater usually extends a single segment of a LAN, to accommodate additional users. The two sides can conceivably use two different media or physical topologies, although this is not a frequent

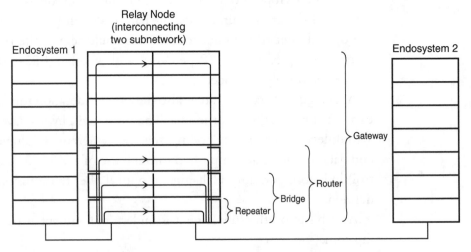

Figure 2.6 Relay functions.

occurrence. Three limitations of repeaters are: (1) the amplification of noise, along with the signal; (2) the limited nature of the extension; and (3) the fact that the network remains a single network at the logical level, thereby keeping the number of users that can be supported bounded by medium-sharing considerations. Repeaters connect LANs at OSI Layer 1.

Bridges[20] connect two or more LANs at the MAC (or LLC) layer. A bridge receiving frames of information will pass the frame to the inter-connected LAN based on some forwarding algorithm selected by the manufacturer (e.g., explicit route, dynamic address filtering, static address filtering, etc.; see Figure 2.7). The protocol layer at which bridging takes place—the data link layer—handles data flow, manages transmission errors, provides physical addressing, and controls access to the physical medium.

---

[20] In principle, there are two kinds of bridges at the protocol level. *MAC bridges* only operate up to the MAC layer; they do not interpret or act upon LLC (data link layer information). This type of bridge is transparent to protocols at and above the LLC. *Data link layer bridges* interpret and act upon information up to and including the LLC; this type of bridge is transparent to protocols at and above the network layer. Most implementations are MAC bridging implementations. Most references to bridges in this book refer to MAC layer bridging.

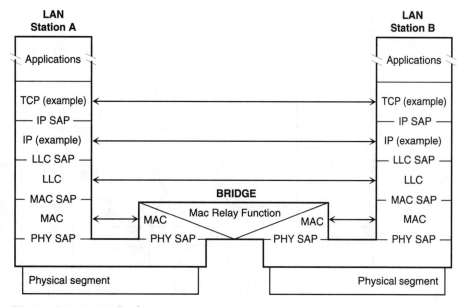

Figure 2.7 *LAN bridge operation.*

Bridges support an appropriate subset of these functions by implementing various link layer protocols' peers in the device.

There are several kinds of bridges. *Transparent bridging* allows forwarding of frames without any intervention of the sender or receiver; in practical terms, this kind of bridging is found in Ethernet environments. *Source-route bridging* requires the specification of the PDU path in order for the data to flow; it is found in Token Ring environments. *Translating bridging* supports translation between formats and transit disciplines of different media (e.g., Ethernet to FDDI or Ethernet to Token Ring; see Figure 2.8). *Source-route transparent bridging* combines the algorithms of transparent bridging and source-route bridging to support communication in heterogeneous Ethernet/Token Ring environments.

From a communication's perspective, bridges can be characterized as either local or remote. As the name implies, *local bridges* connect LAN segments in the same geographic proximity, without an intervening telecommunication link; this type of bridging is normally at full LAN speed. *Remote bridges* connect segments that are not in geographic proximity. Remote bridges face two challenges: (1) The protocol architectures of the

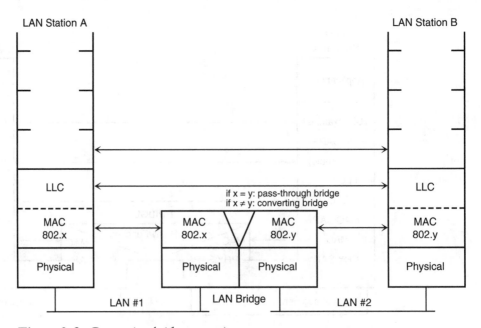

Figure 2.8 Converting bridge operation.

public network (e.g., physical-level T1/T3 lines, X.25 packet-switched networks, frame relay networks, and ATM networks) are not the same as the architecture of a LAN, making protocol adaptation necessary;[21,22] and (2) the WAN speed is rarely comparable to that of the LAN, making some kind of compromise necessary—this becomes an issue in many delay-sensitive applications. Remote bridges compensate for speed differences by using adequate amounts of buffering. WAN interconnection is now more often supported via routers than via bridging.

Routers connect at a higher protocol layer (the network layer) than bridges. Routers provide flow control for the incoming LAN packets, thereby increasing the reliability of the interconnection, and allow the use of a variety of interconnection subnetworks. Different network layer PDUs can, in principle, be routed over different networks; for example, for security or least-cost routing reasons. Routers operate with a particular WAN protocol or with a number of protocols. Routers typically need to support a number of WAN technologies (e.g., private lines, frame relay, cell relay, etc.);[23] they face the same restrictions faced by bridges in terms of WAN protocol architecture and communication speed. If multiple protocols are being used to interconnect LANs, a manager can either select a separate router for each protocol or have a router that is capable of retaining multiple protocols in one chassis. Disadvantages of routers, relative to bridges, include lower packet filtering speed (due to increased processing requirements) and increased cost, although differences with reference to bridges are becoming smaller.

Gateways are used to interconnect LANs (in general, all subnetwork technology) that employ completely different protocols at all communication layers. The complete translation of incoming data units associated with completely different protocols affects transmission speed. Gateways connect LANs at Layers 4 through 7 of the OSI Reference Model.

The rest of the chapter focuses on bridging technology in support of L2S techniques. In effect, L2Ss are multipoint bridges.

---

[21] In jargon, one *tunnels* (encapsulates) IEEE 802 frames in WAN frames.

[22] ATM will resolve this problem when both the LAN and the WAN have migrated to this technology—this will occur during the next few years.

[23] They can also support LAN protocols (e.g., 802.3 or FDD), when used to connect subnetworks in intracampus or intrabuilding applications.

Functions of a Bridge

Bridges are basically simple devices. They accept incoming LAN frames, make forwarding decisions based on information contained in the frames, and forward the frames along to their intended destinations in the connected subnetwork; frames are forwarded one hop at a time to their destinations.[24] Table 2.2 describes the advantages of bridging; these advantages originate from partitioning traditional shared-medium LANs into smaller subnetworks (generally, there are no comparable advantages in non-shared-medium LANs, since these do not require segmentation to manage performance). Vendors have added features to bridges over time, including large address tables, complex frame filtering, increased throughput, load balancing, a wider selection of WAN interfaces, redundancy, and support of network management capabilities via SNMP. A number of vendors have also added variable levels of routing, allowing users to deploy bridges to interconnect two (or more) LANs and later invoke routing capabilities as additional protocols are added to the network. Figure 2.9 depicts some of the internals of a bridge, at the logical level.

[24] In source-route bridging the entire path to the destination device is described in the frame.

Table 2.2 Advantages of Bridging

ADVANTAGE	DESCRIPTION
Reduced traffic	Only a fraction of the traffic is forwarded; decreased contention traffic is experienced by all connected subnetworks.
Firewalling	Bridges offer MAC-level firewalls (filtering based on policies related to MAC addresses), screening some devices from potential infractions generated by other devices.
Increased population	Interconnection of subnetworks of discrete broadcast/token domain increases the population of devices able to be connected.
Increased geographic reach	Allows users to connect at distances exceeding the LAN limits, whether local or off-campus, and be interconnected.

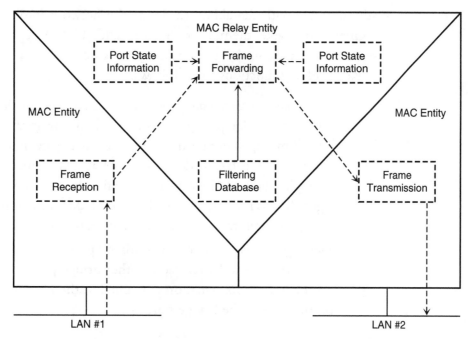

Figure 2.9 *Functional view of a bridge.*

As noted, a bridge interconnects multiple LANs, providing communications between them, while isolating traffic where appropriate. A bridge (regardless of the type) examines the destination address of each MAC PDU it receives: If the source and destination of the PDU are on the same LAN segment, the bridge discards the PDU; on the other hand, if the destination is on a different LAN segment, the bridge forwards the PDU. The bridge learns on which port the destination can be reached. If the destination is not in the address table, then all ports are flooded; thereafter, the bridge watches to see on which port a response to the original PDU is received. This information is placed in the bridging table. The next time that a PDU is received for the destination in question, only the appropriate port receives a copy of the PDU for transmission. The receiving LAN must typically run the same MAC protocol as the transmitting LAN in order to read the PDU. (Although translating bridges are also available for this function, many designers now use routers instead.) As networks become complex due to the addition of multiple departments and additional servers that share a common backbone, bridges provide the network administrator with the ability to divide the network into smaller logical segments to make them more manageable. Unlike repeaters, bridges regenerate signals, so that noise is not propagated.

Key bridging functions are as follows:

- *Maintenance of port state information.* State information associated with each bridge port governs its participation in the bridged system. If management permits a port to participate in frame relaying, and if it is capable of doing so, then it is described as *active*. For example, the 802.1D standard specifies the use of a spanning tree algorithm (STA) and protocol, which reduces the topology of the bridged system to a simply connected active topology.
- *Frame forwarding.* The forwarding process forwards received frames that are to be relayed to other bridge ports, filtering frames on the basis of information contained in the filtering database and on the state of the bridge ports.

Bridge Categories

LAN bridges can be classified as pass-through and converting (also known as *translation bridges*). *Pass-through bridges* connect similar LANs and provide such functions as repeating, frame filtering, and frame queuing; frames are passed without any MAC conversion. *Converting bridges* interconnect dissimilar LANs and must, therefore, convert the header and trailer of MAC incoming frames. See Table 2.3 for a characterization of different bridge types. This topic is revisited in the next subsection.

Segment-to-segment bridging is relatively straightforward. Multiple cascaded segments are also possible; however, the traffic destined for a "far" node must pass through several bridges, causing a possible degradation in quality of service (increased delay and increased probability of frame loss). Multiple LANs often use an interconnecting backbone to avoid cascading and local traffic congestion (as discussed, the LAN frame may be encapsulated and/or segmented via the backbone protocol). FDDI or ATM are typical backbone technologies. Multiport bridging allows several LANs to share the bridge. Here, devices communicate with each other through the bridge's internal bus. Shared resources and improved network management can reduce the cost per port of bridging. By incorporating a physi-

Table 2.3 Five General Categories of Bridges

CATEGORY	DESCRIPTION
Transparent bridges (TBs)	The terminology *transparent* originates from the fact that end systems are not aware of the bridge or of the fact that there is more than one LAN segment. Transparent bridges look up the address of each PDU in a routing table, which the bridge compiles from observing the addresses of PDUs originating on a segment and passing on those which do not match (being, therefore, destined for a node on a different segment).
Encapsulating bridges	These bridges use an encapsulation method to bridge two identical MACs via a different type of medium. This kind of bridging allows an FDDI ring to act as a backbone for the interconnection of several bridged Ethernet LANs.
Translation bridges	Used, for example, when devices on an Ethernet network need to exchange data with devices on an FDDI LAN (this could be handled by a router). The translation bridge "translates" the Ethernet MAC frames and the MAC-layer address into items conformant to FDDI. The translation bridge also fragments FDDI frames, if necessary, before forwarding them on an Ethernet LAN.
Source-routing bridges	These are employed in IBM Token Ring LAN. They utilize the routing information field (RIF) in the MAC header of a Token Ring PDU to determine on which ring(s) the PDU should be transmitted.
Source-routing transparent (SRT) bridges	These bridges are a combination of source-routing and transparent bridging. On receipt of a PDU the bridge examines the PDU to determine if a routing information indicator (RII) is set. If so, the RII is followed by a RIF; the bridge then treats the PDU as a source-routing PDU. If the RII is not set (and the RIF is not present), the bridge treats the PDU as a transparent-bridging PDU. These bridges allow IBM and non-IBM nodes to share the same LAN.

cal layer relay function, bridges can also interconnect different physical media types, such as 10Base-T and fiberoptic cable. The interconnected LANs can also utilize differing access methods, such as Ethernet and FDDI, by using a translation bridge.

MAC-layer bridges operate in a manner that is independent of the protocols above the MAC layer; that is to say, the decision to keep traffic local or to forward the MAC PDU is made *without* consideration of upper-layer protocols encapsulated inside the MAC frame. A *MAC bridge* operates below the MAC service boundary and is transparent to protocols operating above this boundary; this includes the LLC sublayer, the network layer, the transport layer, and any other upper layer that may be employed by the user's application. Bridges, therefore, have been considered a reasonable interconnect solution for physically homogeneous (i.e., same MAC) but protocol-heterogeneous network environments. Another way of looking at this is that since bridges operate at Layer 2, they are not required to analyze upper-layer information. Because the forwarding decisions are based only on the outmost MAC protocol control information (PCI)—specifically, the destination address—the forwarding decisions are fast. Hence, these bridges can be forwarding frames containing AppleTalk, DECnet, TCP/IP, SNA, and XNS traffic between networks. However, at some point, some entity in the network will have to reconcile the network layer differences.

By "unwrapping" Layer 2 PCI, *data link layer bridges* are capable of filtering frames based on any data link layer information. (Note that the data link layer includes both MAC and LLC.) For example, a bridge can be configured to drop (i.e., not forward) all frames originating from a particular subnetwork. Since data link layer PCI contains a reference to the kind of upper-layer protocols being *carried*—that is, the SNAP PID—the bridge can usually filter on this parameter (although it does not *interpret* the higher-layer protocols). Filters are used to control broadcast domains and multicast PDUs.

It is conceivable that the interconnected LANs use different maximum PDU sizes; hence, it is important that the higher-layer protocols encapsulated in MAC frames passing through the bridge do not violate the maximum PDU size for any LAN segment. (A bridge cannot correct the discrepancy since it would require knowledge of the network layer in use, which is something it does not support.) This segmentation must usually be handled by the upper-layer protocols (e.g., IP and/or TCP).

Technical Details on Bridge Technology

This section provides additional details on bridging technology. Although pure bridging is no longer used as a primary standalone technology, its basic concept lives on in L2Ss and in ATM LANE. VLANs can also utilize bridging.

Transparent Bridging (TB)

Transparent bridges (TBs) have been a common technology on Ethernet LANs. The presence and operation of TBs is transparent to network devices. Transparent bridges are bridges that do not require the user to specify the path to the destination. The TB's address table is maintained on a static or dynamic basis. In the static environment, the LAN administrator specifies whether or not a frame for a given destination needs to be forwarded to the downstream LAN. In a dynamic environment, the bridge builds its own table by *observation*. Most bridges now use dynamic techniques. Every bridge in the network must maintain a table with entries for all active users. Because of table entry limitations, the bridges employ aging techniques to delete destination entries having no recent activity. If a frame arrives at the bridge for a destination that is not on the table, a flooding technique is used, as explained earlier; however, this impacts the overall performance of the bridge. Hence, a larger address table is advisable. Frame filtering and forwarding rates vary from a few thousand per second at the low end, to a few tens of thousands at the midrange, to a few hundreds of thousands at the high end. Rates in the 10,000 to 20,000 range are common.

TBs utilized in Ethernet need to maintain address tables. Until the early 1990s, these tables were able to store just a few thousand entries; newer bridges can store up to 100,000 entries (although tables of 20,000 to 50,000 entries are more typical). In contrast, *source-routing* bridges used in Token Ring LANs require the sending user station (i.e., the source) to supply instructions as to how to reach the destination. TBs were first introduced by Digital Equipment Corporation in the early 1980s; these concepts were subsequently incorporated into IEEE standards.

When they are powered up, TBs learn the network's logical topology by analyzing the source MAC addresses of incoming frames from all devices that are connected to the bridge. For example, if a bridge receives a frame on port 1 from device A, the bridge infers that device A can be reached through the network connected to port 1. A TB can then populate a table showing MAC device addresses and network and port numbers.

TBs isolate intrasegment traffic from leaving the segment, thereby reducing the propagation of frames to other segments. In turn, this reduces the rate of collisions and transported traffic and improves overall performance. Performance depends on the amount of intersegment traffic compared to the total traffic, including broadcast and multicast traffic.

A TB uses the internal table as the mechanism for traffic forwarding. When a frame arrives on one of the TB's ports, the TB looks for the frame's destination address in the table. If an entry is found showing the destination address and one of the TB's ports (other than the one on which the frame was received), the frame is forwarded to that port. If no entry is found, the frame is flooded to all ports except the one on which the frame originally arrived. Similarly, frames requiring broadcast and multicast are also sent to all ports except the one on which they arrived.

Multiple paths between bridges are desirable in principle, because they increase the reliability of the overall connection. However, problems arise when there is no bridge-to-bridge protocol to control loops. Consider as an example user A connected to network 2 and user B connected to network 1. Assume that there are two bridges (bridge A and bridge B) connecting these two LANs. Assume now that user A sends a MAC frame to user B. Both bridges correctly determine that user A is on network 2. However, because all users receive all frames on broadcast LANs, after user B receives two copies of user A's MAC frames, both bridges will again receive the frame on their network 1 port. It is conceivable that the bridges will then change their tables to show that user A is on network 1. Then, when user B replies to user A's data transaction, both bridges will receive and then discard the reply information because their tables show that the destination (user *A*) is on the same segment as the source user. Another problem that arises is the proliferation of broadcast messages, leading to severe performance problems.

The IEEE has published the IEEE 802.1D standard defining an architecture for the interconnection of IEEE 802-based LANs using TBs. As per all MAC layer bridges, the interconnection is achieved below the MAC layer and is, therefore, transparent to the LLC and all upper layers. An IEEE 802.1D bridge operation is such that it provides redundant paths between end stations to enable the bridged LANs to continue to provide the service in the event of component failure (of bridge or LAN). The standard includes the Spanning Tree Algorithm (STA), which ensures a loop-free topology while providing redundancy. Features supported by IEEE 802.1D bridges include the following:

1. A bridge is not directly addressed by communicating end stations: Frames transmitted between end stations carry the MAC address of the peer-end station in their destination address fields, not the MAC address of the bridge (if it has any). The only time a bridge is directly addressed is as an end station for management purposes.

2. All MAC addresses must be unique and addressable within the bridged system.

3. The MAC addresses of end stations are not restricted by the topology and configuration of the bridged system.

4. The quality of the MAC service supported by a bridge is comparable, by design, to that provided by a single LAN. Key aspects of service quality are:

   • Service availability
   • Frame loss
   • Frame misordering
   • Frame duplication
   • Transit delay experienced by frames
   • Frame lifetime

- Undetected frame error rate
- Maximum service data unit size supported
- User priority
- Throughput

An 802.1D bridge relays individual MAC user data frames between the separate MACs of the LANs connected to its ports. The order of frames of given user priority received on one bridge port and transmitted on another port is preserved. The functions that support the relaying of frames are the following:

1. Frame reception
2. Discard of the frames if received in error
3. Frame discard if the frame type is not user data frame
4. Fame discard following the application of filtering information
5. Frame discard if service data unit size is exceeded
6. Forwarding of received frames to other bridge ports
7. Frame discard to ensure that a maximum bridge transit delay is not exceeded
8. Selection of outbound access priority
9. Mapping of service data units and recalculation of frame check sequence
10. Frame transmission

Specific to the STA, a bridge filters frames in order to prevent unnecessary duplication. For example, frames received at a port are not copied to the same port. Frames transmitted between a pair of end stations are confined to LANs that form a path between those end stations. The functions that support the use and maintenance of filtering information include the following [1,2]:

1. Automatic learning of dynamic filtering information through observation of bridged system traffic
2. Aging of filtering information that has been automatically learned
3. Calculation and configuration of bridged system topology

The STA was designed to retain the fault–tolerance benefits of physical loops, while eliminating the possible side effects. The algorithm was initially advanced by Digital Equipment Corporation (Ph.D. thesis of Dr. R. Perlman [7]) and later became an IEEE standard. The STA creates a loop-free topology by selecting a subset of paths. This is done by placing those bridge ports that would create loops if activated into a standby condition. In normal operation, these bridges do not forward onto these standby

2	1	1	1	8	4	8	2	2	2	2	2
Protocol identifier	Version	Message type	Flags	Root ID	Root path cost	Bridge ID	Port ID	Message age	Maximum age	Hello time	Forward delay

Note: All lengths in bytes.

Protocol identifier	Contains the value 0.
Version	Contains the value 0.
Message type	Contains the value 0.
Flags	TC bit (bit 1) signals a topology change. TCA bit (bit 2) is set to acknowledge receipt of a configuration message with the TC bit set. Bits 3–7 are unused.
Root ID	Contains the identity of the root bridge (2-byte priority followed by 6-byte ID).
Root path cost	Cost of path from the bridge sending the configuration message to the root bridge.
Bridge ID	Priority and ID of the bridge sending the message.
Port ID	Indentity of port from which the configuration message was sent (used for loop identification).
Message age	Amount of time since the root sent the configuration message upon which the current configuration message is based.
Maximum age	Indicates time when current configuration message should be deleted.
Hello time	Time period between root bridge configuration messages.
Forward delay	Length of time that a bridge waits before transitioning to a new state after a topology change. (If time is too short, loops may result.)

Figure 2.10 *Configuration message format for TBs.*

ports. However, blocked bridge ports can be activated in the event of link failure on the primary path. This, in turn, provides a new path through the internetwork apparatus. The STA algorithm is based on mathematical theorems that indicate that for any graph there is a *spanning tree* (a tree that connects all the nodes). By definition, a *tree* does not contain any loops.

The STA calculation takes place when the bridge is turned on and whenever a change in the topology is detected. The computations require communication between TBs. This is accomplished through *bridge protocol data units* (BPDUs), also called *configuration messages*. BPDUs contain information flagging the bridge that is presumed to be the *root* (root identifier) and the *root path cost* (i.e., the distance) of the sending bridge (see Figure 2.10). BPDUs also contain the bridge and port identifier of the sending bridge, along with age information. Bridges exchange BPDUs at short intervals (typically, 1 to 4 s). All the topology decisions are made in a distributed fashion (locally): BPDUs are exchanged between neighbor bridges, and there is no central control on the topology. If a bridge experiences an outage and a topology alteration ensues, neighboring bridges will detect the lack of reception of BPDUs and thereby initiate a STA recalculation. Topological change messages are shown in Figure 2.11.

Source–Route Bridging (SRB)

Source-route bridging (SRB) was initially proposed as a way to perform generic LAN bridging. However, with the acceptance of STA in IEEE

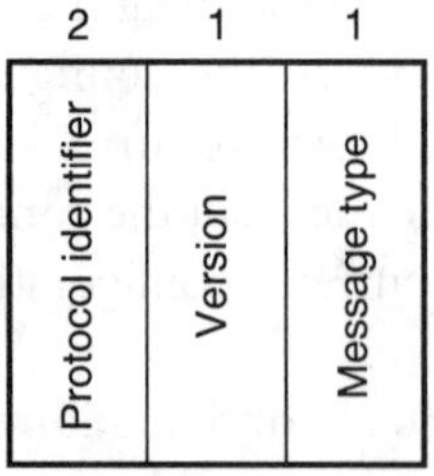

2	1	1
Protocol identifier	Version	Message type

Note: All lengths in bytes.

Protocol identifier   Contains the value 0
Version   Contains the value 0
Message type   Contains the value 128

Figure 2.11 Topological change messages.

802.1D, SRB found its way into the IEEE 802.5 as a way to bridge Token Ring LANs. More recently, IBM has proposed source-route transparent (SRT) bridging; this technology eliminates the need for pure SRBs. This means that TBs and SRTs will be the important technologies from this point forward, although SRBs are still to be found in a number of enterprise networks.

SRBs get their name from the fact that the complete source-to-destination route is shown in all inter-LAN MAC frames. The route information is contained in the routing information field (RIF; see Figure 2.12).

Consider the LAN internetwork shown in Figure 2.13. Assume that end station x needs to send information to end station y. At startup, end station x does not know whether end station y is on the same segment. To establish this, end station x issues a *test frame:* If the frame returns to end station x without an affirmative indication that end station y has seen it, end station x must assume that end station y is on a remote segment. Next, end station x must determine the exact location and route of end station $y;$ this is accomplished by sending an *explorer frame.* Each bridge receiving the explorer frame transmits the frame onto all outbound ports. Route information is added to the explorer frames as they journey through the interconnecting links. When end station x's explorer frames reach end station y, end station y responds to each frame individually, utilizing the route information that has accumulated in the frame. Finally, when end station x receives all of the response frames, it chooses a path based on some choice criterion.[25] After a route is selected, it is inserted into the MAC frames for end station y in the RIF (the presence of the RIF is signaled by setting the most significant bit of the *source* address field, called the *routing information indicator* [RII]). In Figure 2.13, bridges 1 and 2 receive explorer frames. The routes that are found are: (1) LAN1/bridge 2/LAN4/bridge 4/LAN2, and (2) LAN1/bridge 1/LAN3/bridge 3/LAN2. One route is selected.

---

[25] The IEEE 802.5 standard does not specify the choice criterion, but the following suggestions are indicated: (1) Use the route of the first frame received, (2) use the route with the minimum number of links, (3) use the route with the largest allowed frame size, or (4) use some combination of the preceding. Other criteria could also be used (e.g., select a route at random, or based on a security discipline, etc.)

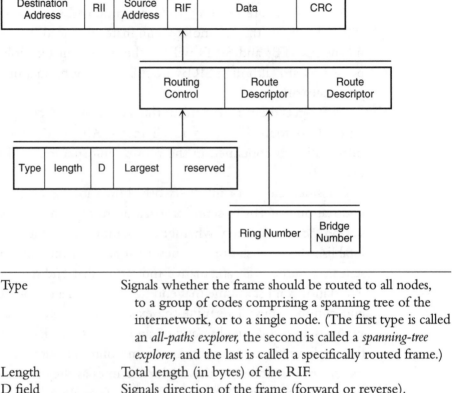

Type		Signals whether the frame should be routed to all nodes, to a group of codes comprising a spanning tree of the internetwork, or to a single node. (The first type is called an *all-paths explorer,* the second is called a *spanning-tree explorer,* and the last is called a specifically routed frame.)
Length		Total length (in bytes) of the RIF.
D field		Signals direction of the frame (forward or reverse).
Largest field		Specifies the largest frame that can be sent along the given route.
Route descriptor field(s)		Each identifies a ring-number pair that specifies a portion of the route. (Routes are alternating sequences of LAN and bridge numbers—sequences start and end with bridge numbers.)

Figure 2.12 A routing information field.

Mixed–Media Bridging

TBs are found in Ethernet networks, while SRBs are typically associated with Token Ring LANs. Since TBs and SRBs are still found in LANs, there is occasionally a need to internet between these islands of networks. *Translating bridging* (TLB) provides an approach to effect this connectivity. TLB entered the market in the late 1980s. There are no standards in support of TLB, so many facets of the interconnection are left up to the implementers.

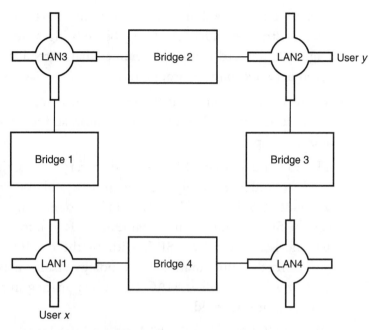

Figure 2.13 *Example of a network with SRB.*

In the early 1990s, IBM addressed some of the weaknesses of TLBs by introducing source-route transparent (SRT) bridging. SRTs can process frames from both source route and transparent bridging systems and form a common spanning tree with TBs. This enables end stations of each type to communicate with end stations of the same type in a network.

Both TLBs and SRTs have limitations. Specifically, there are technical challenges associated with enabling stations from a SRB/Token Ring subnetwork to interwork with an Ethernet/TB subnetwork, including, but not limited to, the following:

- Although both Ethernet and Token Ring use a 48-bit MAC address, the bit ordering is different. When looking at the serial bit representation of the address, Token Ring assigns the first bit encountered in the stream to be the high-order bit of the byte, while Ethernet considers such a bit to be the low-order bit.

- In some cases MAC addresses are actually carried in the data file of the frame (e.g., in the Address Resolution Protocol [ARP]). Conversion of these kinds of addresses as the frame goes across the internet is difficult.

- There are incompatibilities in the maximum transfer unit (MTU) sizes of Ethernet and Token Ring. Since bridges are not capable of fragmentation and reassembly, frames that exceed the MTU of a given network must be dropped.

- Status bits of Token Ring frames are not mappable to Ethernet. Other functions, such as priority, monitor, and reservation bits, are also not mappable.

- TBs do not understand SRB route discovery frames, since they discover topology by analyzing the source addresses of incoming frames. Related to this, the RIF field has no meaning to TBs. Conversely, SRBs require that all frames have RIF information. When a frame is received by a SRB without the the RIF information (including the configuration and topology change messages generated by a TB, as well as MAC frames originating in the TB environment), it is ignored.

- SRB and TB both use the STA to eliminate loops, but the implementations of the algorithms are incompatible. Hence, multiple paths between the SRB domain and the TB domain are not supported.

Translating Bridging (TLB)

Given the lack of standardization of TLB, there are several variants of implementation. The approximate operation of a translating bridge (TLB) is as follows:

1. Reorder source and destination address bits.

2. Deal with embedded MAC addresses by programming the bridge to scan for various types of MAC addresses. (This solution must be adapted with each new type of embedded MAC address.) Some bridges only check for the most popular embedded addresses.[26]

3. Set MTU to 1500 in the TLBs that forward frames from the TB environment to the SRB environment. The RIF field has a subfield that indicates the largest MAC frames that can be accepted by

---

[26] In those cases where the TLB is embedded in a multiprotocol router, the router can route these protocols, thereby eliminating the need for the bridge to have to address them.

the SRB. (This feature is implementation-dependent.) In those cases where the sending device on the SRB side cannot respond to the MTU negotiation from the TLB, the TLB is forced to drop frames that exceed the 1500 value of MTU for Ethernet. In those cases where the TLB is embedded in a multiprotocol router, the router can route these protocols, thereby eliminating the need for the bridge to have to address them.

4. PCI information representing Token Ring functions that have no Ethernet equivalent (e.g., priority, reservation, and monitor[27]) is ignored by TLBs.

5. When the bridged traffic arrives from the SRB domain destined for the TB domain, the SRB PCI is removed. RIFs may be cached for use by frames representing return traffic.

6. When the bridged traffic arrives from the TB domain destined for the SRB domain, the TLB checks to see if the frame has a specific destination address. In that case the TLB looks up the destination in the RIF cache; if a path is identified, it will be used by adding the RIF information to the frame. If no entry is found in the cache, the frame is sent to the spanning tree explorer. If the frame has a multicast or broadcast destination it is sent to the spanning tree explorer, which will appropriately handle the frame.

In effect, the TLBs create a software (Level 2) gateway between the two domains. To a SRB station, the TLB has a bridge number and a ring number associated with it that reflect the entire TB domain. Hence, the TLB looks like a standard SRB. To the TB domain, the TLB looks like a TB.

Source-Route Transparent (SRT) Bridging

Source-route transparent bridges (SRTs) combine the TB and SRB algorithms. However, the translation is not perfect, because of the MAC protocol incompatibilities previously listed. SRTs employ the RII bit to

---

[27] Token Ring status bits are treated in vendor-specific ways. Some vendors ignore them. Other vendors set the C bit (copied the frame), but not the A bit (destination seen the frame). In the former case, it is impossible for a Token Ring sending station to determine if the frame was lost; responsibility is pushed up the protocol stack (e.g., in TCP/ISO TP).

distinguish between frames that use TB and frames that use SRB. If RII = 1 a RIF frame is present, and the bridge uses the SRB algorithm; otherwise, the bridge uses TB. SRTs are typically implemented as hardware upgrades to a SRB (in addition to software upgrades) to enable them to analyze every frame.

Virtual LANs (VLANs)

VLANs allow the logical addresses of clients and servers to be separated from their physical locations. In general, LAN ports are administratively configured to appear to all come from one common hub. Clients are manually assigned to a VLAN, and the ports that make up the virtual LAN are kept in a database. This model operates as a MAC layer bridge transporting messages between members of the virtual LAN. Moving a client from one port to another removes it from its virtual LAN, requiring managed virtual LAN databases to be manually updated to reflect these changes. Both the enterprisewide network evolution and the VLAN problem benefit from ATM technology. Hence, a mechanism to make effective use of ATM capabilities is sought. LANE (see Chapter 4) is both a bridging- and switching-based technology that utilizes ATM to support emulated LANs (that is, VLANs). This approach will also make VLANs standards-based.

Until the emergence of LANE (1996 to 1997) and MPOA (1997 to 1998), VLAN capabilities tended to be primitive in both Ethernet and Token Ring switches. In the Token Ring context, *VLAN switching* refers to the ability to group not only a set of ports on a switch but also a set of ring numbers, in such a way that traffic on the selected ports or rings, including broadcasts, is never forwarded to foreign ports or rings. More advanced VLAN functions (e.g., the ability to associate a set of users based on their organizational or functional responsibilities irrespective of topology) require advanced software, Layer 3 protocol capabilities, and interoperability mechanisms that encompass hubs, switches, routers, and backbone (ATM) devices.

Disadvantages and Limitations of Bridge-Based Interconnection

A bridge-based internetwork philosophy implemented on a large scale has several disadvantages, including the following:

1. Difficulty in mixing LAN technologies—for example, Ethernet and Token Ring.
2. Difficulty in controlling broadcast storms
3. Lack of firewalling for application and security reasons
4. Inability to select traffic routes based on the kind of networking protocol
5. Performance problems related to treatment of traffic upon bridge congestion
6. Limited WAN connectivity

These and other limitations drive network planners to other technologies, such as routers and MPOA.

LAN Switching

As previously discussed,[28] for the past few years network managers have had several tools for building large, complex networks. This tool kit has consisted of repeaters, hubs, bridges, routers, and gateways. Recently, a new tool, the *LAN switch,* has emerged. The impact of LAN switching has been significant, causing network administrators to rethink the fundamental rules of network design, and leading to questions concerning the function and placement of switches and routers in corporate networks. This discussion helps identify the capabilities and applications of switches and routers so that network designers can optimize their networks.

A *switch* is a special-purpose device specifically designed to address LAN performance problems resulting from bandwidth limitations and network bottlenecks. A switch economically segments the network into smaller collision domains, providing a higher percentage of bandwidth to each end station. The use of application-specific integrated circuit (ASIC) technology allows the switch to deliver higher performance over all ports at a relatively low cost per port.

[28] Some portions of this material were provided by Dana Christensen, David Flynn, and Chuck Semeria of 3Com, Santa Clara, Calif. The material is used with permission of 3Com. 3Com provides a complete family of workgroup hubs, switches, and NICs, in addition to many other products.

A *router* is a general-purpose device designed to segment a network with the goals of limiting broadcast traffic and providing security, control, and redundancy between individual broadcast domains. As previously discussed, a router operates at a higher layer of the OSI Reference Model, distinguishing among network layer protocols and making intelligent packet-forwarding decisions. A router also provides firewall service and WAN access.

If a network application requires limiting broadcast traffic, support for redundant paths, intelligent packet-forwarding, or WAN access, a router must be selected. If the application requires only increased bandwidth to ease a traffic bottleneck, a switch is possibly the better choice. The technology choices appropriate for a specific workgroup, department, or building backbone depend upon the organization's business and technical requirements.

Switching and routing are complementary technologies that allow networks to scale to sizes beyond those that can be achieved using either technology alone. Rather than choosing between switching and routing, network designers can combine the two technologies to build high-performance scalable networks.

The best way to solve internetworking problems is to look at the business and technical requirements of the network, and then deploy the technologies that achieve the organization's specific design goals in the most cost-effective manner. Any proposed solution must carefully consider application needs, traffic patterns, and workgroup composition. The solution must also consider the capabilities of the existing network, security, scalability, ease of use, and network management.

Most environments will likely use both switching and routing. In certain applications within a network, a switch provides the right solution. In others, a router is the more appropriate choice. And there are some applications for which either a switch or a router may provide a satisfactory solution, based on a careful examination of the network design goals.

Motivations for Switched LANs

With the rapid increase in processing power of workstations and servers implied in the previous chapter, the proliferation of bandwidth-hungry applications, and the growth of local networks, many network managers are facing an urgent need for more bandwidth and increased workgroup performance. The need for more bandwidth is a common theme, whether the workgroup is a large and growing LAN with hundreds of users running e-mail and office productivity applications; a medium-sized LAN needing bandwidth for videoconferencing, learning applications, and World Wide Web (WWW) access; a small group of power users running CAD and graphics applications; or even a far-flung collection of remote offices sharing data with one or more central sites.

As discussed in Chapter 1, two approaches to scaling performance in Ethernet networks are receiving a lot of attention: switching and fast Ethernet. Many network managers are wondering which is best for their network. In fact, it is not a matter of choosing one or the other. Switching and fast Ethernet are complementary technologies for the workgroup, and many solutions will include both. The challenge is to figure out just where to deploy each one in order to realize the maximum benefit without unnecessary cost.

Changing Needs in the Workgoup

Over the last ten years, Ethernet LANs have typically been built utilizing shared 10-Mbps technology, and to date this infrastructure has served the majority of customers very well. When congestion has been an issue, the focus on relieving performance bottlenecks has been on the backbone. However, advances in the performance and complexity of PCs, servers, and applications are requiring network managers to scale the performance of their networks all the way to the desktop.

With the arrival of such high-performance microprocessors as the Pentium and PowerPC, advanced disk storage methods, and rapidly decreasing processing costs, the way is paved for powerful, mission-critical, PC-based LAN applications that until recently would have been possible only on a mainframe. Emerging data-intensive applications and technologies include multimedia, groupware, imaging, WWW, intranets, and client/server databases.

Changes in the placement of servers in relation to the workgroup are also affecting workgroup bandwidth requirements. In recent years, there has been a trend toward the centralized placement of servers to simplify server maintenance and improve network security. Many organizations have also added high-performance servers to consolidate multiple workgroup applications. These centralized servers require more bandwidth to serve power users and traffic from multiple workgroups.

Until recently, the cost of implementing a high-speed solution to the desktop was too high to be considered for widespread deployment, but switched Ethernet, Fast Ethernet NIC, and shared-media hub prices have decreased significantly. The combination of high-performance end systems and low-cost, high-performance networking solutions is leading many network managers to reevaluate their workgroup design strategies.

Determining the Need for Increased Bandwidth

To keep up with changing workgroup bandwidth needs, network administrators must assess the current utilization of their networks, as well as plan for future growth and application requirements. This assessment involves answering questions like the following:

- Are users currently complaining about poor response times?
- What is the average utilization of the network?
- What is the plan for network expansion?
- What applications are currently running?
- What new applications are expected to be deployed over the next three to five years?
- Where are servers located? Do they typically serve one workgroup, or do they serve users from multiple workgroups?
- How easy is the network to manage and keep operational?

Network management software plays a key role in determining the current utilization of the network. Network management software and built-in Remote Monitoring (RMON) management in hubs and switches can be used to set thresholds and gather network statistics. This can provide important trend information and can alert the network manager to the need for increased bandwidth. When average network utilization is between 30 and 40 percent (assuming other possible causes for poor performance have been ruled out), more bandwidth is needed.

To maintain competitiveness, many organizations use a three-year planning cycle for PC upgrades. To match this cycle, network managers can use a three- to five-year planning horizon for the introduction of new applications and network infrastructure requirements. A predictive performance model can help provide a "what-if" estimate of network utilization levels and end-user response times.

Switching Technology

Switching technology operates at Layer 2 of the OSI Reference Model. The emerging popularity of switching products can be viewed as a resurgence of bridge technology in a simpler, lower-cost, higher-performance, and higher-port-density device. Like a bridge, a switch makes a relatively simple forwarding decision based on the destination MAC address contained in each packet. Generally, this forwarding decision does not consider other information buried deep inside the packet. Unlike a bridge, a switch can forward data with very low latency, providing performance that is closer to single-LAN performance than bridged-LAN performance.

Switching technology allows bandwidth to be scaled in both shared and dedicated LAN segments and can alleviate traffic bottlenecks between LANs. Today, switching products are available for Ethernet, Fast Ethernet, FDDI, Token Ring, and ATM technologies.

Like traditional bridges, switches provide many internetworking benefits. Switches economically segment the network into smaller collision domains, providing a higher percentage of bandwidth to each end station. Their protocol transparency allows them to be installed in networks running multiple protocols with little or no software configuration. Switches use the existing cable plant and station adapters without expensive hardware upgrades. Finally, their total transparency to end stations makes administrative overhead very low, simplifying adds, moves, and changes.

In addition to these benefits, the use of ASIC technology allows a switch to provide greater performance than a traditional bridge by supplying high packet throughput with extremely low latency. This allows a switch to simultaneously forward packets across all ports at wire speed. For example, a single Ethernet interface can support a maximum theoretical transmission of 14,880 packets per second (PPS) for 64-octet (minimum-size) frames. This means that a 12-port wire-speed Ethernet switch, supporting 6 concurrent streams, must provide an aggregate throughput of 89,280 PPS (6 streams × 14,880 PPS/stream). The use of ASIC technology allows the switch to deliver this performance over more ports and at a lower cost per port than a traditional bridge.

Forwarding Models

Switches forward traffic based on one of two forwarding models: cut-through switching and store-and-forward switching.

Cut-through switches start the forwarding process before the entire frame is received. Since the switch only has to read the destination MAC address before it begins to forward a frame, packets are processed faster and latency is at the same low level for both short and long packets. The major disadvantage of pure cut-through switching is that corrupted frames, such as runt packets, jabber packets, and frames with frame check sequence (FCS) errors, are forwarded by the switch. Cut-through switching brings the most benefit when traffic is switched between ports that have the same LAN speed. In contrast, a packet traveling from a 100-Mbps port to a 10-Mbps port will experience some level of buffering as the packet is forwarded.

A *store-and-forward switch* reads and validates the entire packet before initiating the forwarding process. This allows the switch to discard corrupted packets and permits the network manager to define custom packet filters to control the flow of traffic through the switch. The disadvantage of store-and-forward switching is that latency increases in proportion to the size of the packet.

Routing Technology Comparison

Routers operate at Layer 3 of the OSI Reference Model and have more software features than a switch. Functioning at a higher layer than a switch, a router distinguishes among the different network layer protocols, such as IP, IPX, AppleTalk, and DECnet. The additional protocol knowledge available to a router allows it to make a more intelligent forwarding decision than a switch.

Like a switch, a router provides users with seamless communication between individual LAN segments. Unlike a switch, a router determines the logical boundaries between groups of network segments. A router provides a firewall service, since it forwards only traffic that is specifically addressed to go across the router. This eliminates the possibility of broadcast storm propagation, the transmission of packets from unsupported protocols, and the transmission of packets destined for unknown networks across the router.

To accomplish its task, a router must perform two basic functions. First, the router is responsible for the creation and maintenance of a routing table for each network layer protocol. These tables may be created either statically, via manual configuration, or dynamically, using a distance-vector or link-state routing protocol. After the routing tables are created, the router is responsible for identifying the protocol contained in each packet, extracting the network layer destination address, and making a forwarding decision based on the data contained in the specific protocol's routing table.

The enhanced intelligence of a router allows it to select the best forwarding path based on several factors rather than just the destination MAC address. These factors can include the hop count, line speed, transmission cost, delay, and traffic conditions. This increased intelligence can also result in enhanced data security, improved bandwidth utilization, and more control over network operations. The disadvantage is that the additional frame processing performed by a router can increase latency, reducing the router's performance when compared to a simpler switch architecture.

Technology Options for Scaling Workgroup Performance

There are four options for increasing Ethernet workgroup performance today:

- *Switched Ethernet (segmented).* A typical workgroup today consists of one or more shared 10-Mbps repeaters. All users and local servers in the workgroup compete for this 10 Mbps of bandwidth. Adding an Ethernet switch to the network and attaching the 10-Mbps repeaters to the switch ports gives each segment 10 Mbps of bandwidth, increasing the effective bandwidth available to the users.

- *Switched Ethernet (private).*   Private Ethernet dedicates 10 Mbps of bandwidth to each end-user. In this environment, the switch takes the place of a repeater for 10-Mbps connection to the desktop and 100-Mbps connection to local servers. Vendors—notably 3Com—have driven the price of switched Ethernet down significantly, offering performance improvements for a slight price premium over repeating technology.

- *Shared fast Ethernet (100-Mbps Ethernet).*   Fast Ethernet is a relatively new technology that increases Ethernet's bandwidth by a factor of 10. Shared fast Ethernet is useful for eliminating the bottlenecks that power users see at the desktop and other users see at file servers located in centralized server farms. It brings a high-speed 100-Mbps pipe to these power desktops and servers at less than twice the connection cost of 10-Mbps Ethernet.

- *Switched fast Ethernet.*   Switched fast Ethernet can be used to relieve bottlenecks in the building backbone or in some small power workgroups. For the next several years, only a small percentage of workgroups will be able to utilize the full bandwidth—or justify the cost—of switched 100-Mbps to the desktop.

Choosing the Right Technology for the Workgroup

The following criteria can help determine which mix of technologies is appropriate for a given workgroup:

- Existing infrastructure
- Current and planned applications
- Traffic patterns and performance requirements in the workgroup
- Manageability and control requirements
- Price sensitivity

Existing Infrastructure

The first step in determining where to deploy switched Ethernet and fast Ethernet is to examine the existing infrastructure and expansion plans of the workgroup. In many environments, the decision will be dictated by the existing cabling and by how much bandwidth the PCs and servers attached to the LAN can accommodate. In other environments, the decision will be based on such factors as traffic patterns and network utilization, performance requirements, manageability and control requirements, and cost. In general, switched Ethernet is an easy way to boost workgroup performance because it requires no changes to existing desktop computers, network adapters, or cabling.

Cabling

One of the first issues to consider when looking for ways to increase performance in the workgroup is the type of cabling installed. Although the deployment of Category 5 cable and fiber has increased dramatically in recent years, Category 3 and Category 4 cabling still represent a large percentage of the installed base. In many of these installations, two pairs of the cable are used for the LAN and the other two pairs for voice or modem applications. Fast Ethernet runs over Category 5 cable using two pairs, and it can be run over Category 3 and Category 4 cable if four cable pairs are available. If only two pairs of Category 3 or 4 are available, switched Ethernet should be used.

Desktop Bus Type

Table 2.4 shows the theoretical and actual bandwidth that can be supported by various bus types. (The actual bandwidth available to a network

Table 2.4 Maximum Throughput by Bus Type

	ISA	PCMCIA	EISA	MCA	PCI
Bus bandwidth (theoretical)	66 Mbps	66 Mbps	264 Mbps	320 Mbps	1056 Mbps
Bus bandwidth (actual)★	10–20 Mbps	10–20 Mbps	64 Mbps	80 Mbps	264 Mbps
Recommended high-speed connection	Switched Ethernet	Switched Ethernet	Switched Ethernet/ Fast Ethernet	Switched Ethernet	Switched Ethernet/ Fast Ethernet

★Based on performance tests in buses with multiple adapters, such as SCSI, IDE, and VGA.

adapter is affected by other adapters, such as SCSI, IDE, and VGA, attached to the bus.) The maximum bandwidth available to an ISA or PCMCIA NIC is between 10 and 20 Mbps; these adapters will never be able to take advantage of 100 Mbps. Although MCA can support upwards of 80 Mbps, no vendor has an MCA Fast Ethernet adapter at press time. Therefore, switched Ethernet is the best way to increase performance in these types of systems.

PCI (and EISA) bus PCs are able to fully utilize fast Ethernet's 100 Mbps of bandwidth, if applications require it. Before deciding which NIC to purchase for new PCI machines, network managers should review their plans for rollout of new applications. The majority of applications that will be deployed over the next several years can be supported by a dedicated 10-Mbps connection to the desktop. If the environment does not require 100 Mbps over the next two to three years, 10-Mbps PCI adapters should be installed. If planned applications will require 100 Mbps, a 10/100 NIC adapter should be installed. Whether these machines should be configured using switched Ethernet or fast Ethernet depends on the applications, traffic patterns, and performance requirements in the workgroup. Many organizations are deploying new PCI machines in 10-Mbps mode, with plans to expand to fast Ethernet in the future.

Server Type

A key part of scaling workgroup performance is eliminating bottlenecks to servers. As with PCs, the solution for eliminating server bottlenecks depends on the server's throughput capacity. Servers can be classified into three basic types: PC desktop servers, PC servers, and superservers.

PC desktop servers have traditionally been standard PCs with the latest processor, a bit of extra RAM, and maybe a larger hard disk or two. They typically reside in a workgroup and serve only that workgroup. Based on either ISA or EISA buses, these servers can often be accommodated by having a direct 10-Mbps connection to a switch.

High-performance *PC servers*—based on EISA or PCI buses, with a Pentium processor, high-performance drive controller, and multiple disk drives—can generally sustain throughput of 25 to 30 Mbps, with peaks of over 80 Mbps. These servers cannot fully utilize dedicated 100-

Mbps, and up to 5 servers per segment or hub can typically be connected cost-effectively with shared fast Ethernet without restricting performance.

Superservers—specially designed servers with PCI buses, multiple processors, and RAID-based disk subsystems—can sustain throughput of over 80 Mbps, so they warrant the highest-performance connections in a network, either to a small shared fast Ethernet backbone segment that serves a small server farm, or directly to a fast Ethernet switch port.

The majority of the installed base does not require greater than shared 100-Mbps today. A small percentage of servers (superservers) can benefit from dedicated 100-Mbps.

Current and Planned Applications

For the next several years, the most widely deployed applications may not require greater than *dedicated* 10-Mbps. Multimedia and digital video, however, may impact this decision. Fortunately, switched Ethernet is also appropriate when the network strategy calls for migration to an ATM backbone. ATM downlink capability on Ethernet switches is becoming available, making it easy to integrate a high-performance Ethernet workgroup into an ATM infrastructure.

Traffic Patterns and Performance Requirements in the Workgroup

Both switched Ethernet and shared fast Ethernet technologies provide significant performance improvements over traditional 10-Mbps Ethernet, but the exact performance level is a function of the number of servers in the workgroup and the number of clients attempting to access the server at any given time.

Single-Server Environment In a single-server environment, the fast Ethernet solution provides higher performance than a 10-Mbps switched environment when only a few clients access the server simultaneously. This is because the fast Ethernet network can provide peak bandwidth of more than 80 Mbps to even the first client to access the server, while each client on the switched Ethernet network can receive a peak bandwidth of

10 Mbps. But with 8 or more clients on either network, the switched 10-Mbps network provides similar performance to the shared fast Ethernet network. At this level of aggregate bandwidth, the network bottleneck shifts from the clients to the 100-Mbps server link, which now exhibits the same loading characteristics in either configuration.

If the performance problem on the LAN is due to a few users occasionally transferring very large files that require peak bandwidth (power workgroups, sophisticated CAD applications, 3D modeling, etc.), shared fast Ethernet to the desktop is the preferred solution. However, if the problem is due to too many users accessing the server, switched Ethernet and shared fast Ethernet will provide about the same aggregate throughput. In this type of environment, the purchasing decision can be based on other considerations, such as management features, security issues, cost, or existing investment in desktop equipment and adapters.

Dual-Server Environment   In a dual-server workgroup (or local server and backbone servers), the switched network has the potential to provide up to twice the throughput of the fast Ethernet network if the two servers are supported on separate switched 100-Mbps ports. Whether the switched network realizes this potential performance advantage is a function of the number of client PCs accessing the server simultaneously. When fewer than 12 users access the servers at once, this configuration is similar in performance to the single-server solution. However, with very high aggregate traffic patterns in the workgroup, the switched 10-Mbps Ethernet solution can provide significant performance benefits if the traffic is evenly balanced between the two servers.

In a dual-server environment with bursty peak traffic, shared 100-Mbps fast Ethernet would be the better solution. But if congestion is due to large workgroups with multiple users putting a steady, uniform load on the network (client/server applications, such as transaction/database processing, document management, video-based training, or videoconferencing), then switched Ethernet should be used.

Manageability and Control Requirements

Unlike repeaters, which forward data to all ports, switches forward data only to the port where the destination node is located. For example, a finance department and a sales department connected to different switch ports will not be able to see each other's data. Switches make LAN traffic more secure while increasing total network throughput.

As noted, VLANs allow network administrators to enhance network performance, improve network security, and reduce the administrative burden of adds, moves, and changes. VLANs allow managers to direct broadcasts only to members of a specific VLAN, even though those members may be spread across multiple switches in the network. The power of VLANs grows as more switching is added to networks, with the highest benefits accruing to networks providing dedicated switched Ethernet ports to each user. Some network designers may choose switched Ethernet over shared fast Ethernet in order to take advantage of the added security of switching and the management control afforded by VLAN capability.

Price Sensitivity

Switches optimized for the workgroup can deliver switched 10-Mbps Ethernet connections at a price per port close to that of repeating hubs. In existing networks, switched Ethernet is often the simplest and least costly performance upgrade available. Switching does not require any changes to the adapters or the cabling system, and there is no need to touch installed PCs.

When new PCs are added to the network, the costs of installing switched Ethernet and fast Ethernet are similar, with switched Ethernet costing slightly less. A fast Ethernet hub port and a desktop switch port cost about the same, while 10/100-Mbps adapters cost about 50 percent more than 10-Mbps adapters. Given the relatively small cost difference, the choice of technology can usually be based on factors other than cost, such as performance requirements, manageability, control, and security.

Decision Tree

Figure 2.14 depicts a decision tree that can be used to select among various LAN switched solutions. This tree does not include desktop ATM, but whenever multimedia is involved ATM should figure in the discussion.[29] In general, the same decision steps can be used to select Token

---

[29] This use of ATM is unrelated to the use of ATM as the campus or WAN backbone.

This decision tree is designed to help you evaluate workgroup technologies based on network status, desktop equipment, applications, performance and management requirements, and existing infrastructure.

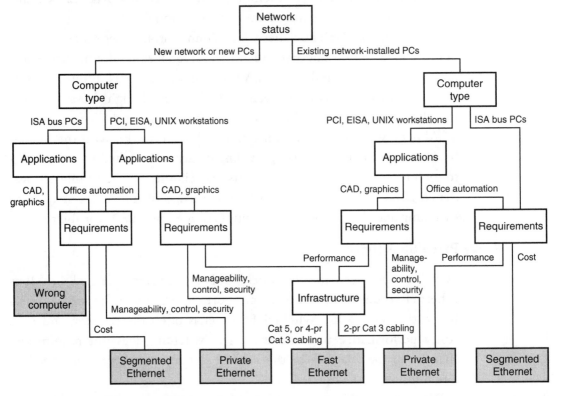

Figure 2.14 *Workgroup technology decision tree.*

Ring versus switched Token Ring, except that in place of "fast" technology, the user could choose 25-Mbps ATM.

Sample Workgroup Configurations

The following workgroup scenarios specifically describe how switched Ethernet and fast Ethernet products help scale the performance of the workgroup. These examples utilize 3Com's family of hubs and switches to support the needed connectivity.

Switched Ethernet with a Shared Fast Ethernet Backbone

The existing network (see Figure 2.15) consists of several 10-Mbps workgroups and a 10-Mbps backbone. Servers had been located within each

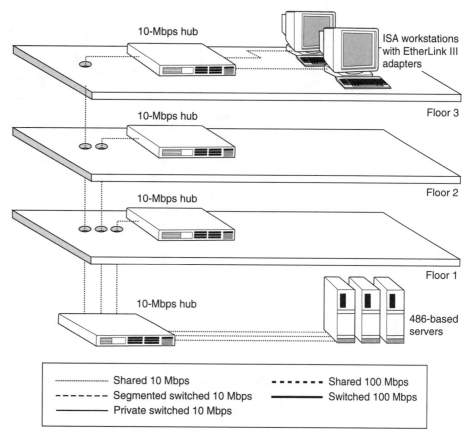

Figure 2.15 *Existing configuration: shared Ethernet.*

workgroup, but were recently moved to a centralized server farm in the computer room. The majority of the traffic is now moving across the backbone, resulting in a significant increase in traffic and a decrease in LAN performance.

- *Installed equipment:* 10-Mbps hubs, ISA workstations with 10-Mbps adapters, and 486-based servers.

- *Expected PC additions:* All new PCs will be Pentium-based with PCI buses.

- *Key applications:* Office applications, including e-mail, word processing, spreadsheet, and presentation packages. New users on Pentium machines will be focused on transaction processing and will require dedicated 10-Mbps.

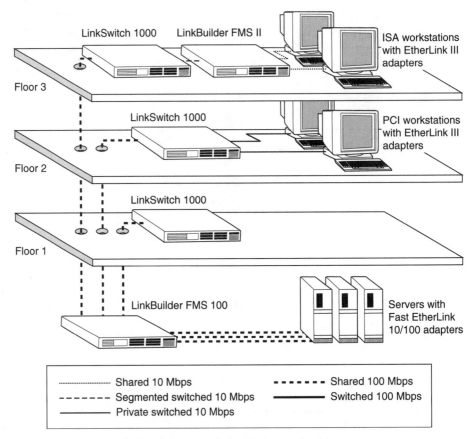

Figure 2.16 *Switched Ethernet with fast Ethernet backbone.*

- *Objectives:* To increase performance in the workgroup, minimizing user response times, and to cost-effectively eliminate bottlenecks in the building backbone.

The solution (see Figure 2.16) relieves server bottlenecks by installing 10/100-Mbps adapters in the servers and connecting them to a fast Ethernet hub. (Because servers are not superservers, they could not take advantage of dedicated 100-Mbps connections.) Ethernet switches provide dedicated 10-Mbps connections to existing Ethernet segments.

Switched Ethernet and Shared Fast Ethernet to the Desktop

The installed base of users with ISA PCs and many of the new PCI bus PCs have begun to deploy graphics-intensive applications and now require dedicated 10-Mbps. In addition, new installations include high-performance UNIX workstations, which will be running CAD/CAM applications and will require greater than 10-Mbps. In addition to the existing 486-based ISA servers, a superserver is being installed to support the new high-performance workgroup.

- *Installed equipment:* 10-Mbps hubs, ISA and PCI PCs with 10-Mbps adapters, and 486-based servers.
- *Expected PC additions:* New UNIX workstations and a new superserver.
- *Key applications:* Graphics-intensive applications and video on PCs and CAD/CAM on UNIX workstations.
- *Objectives:* To scale desktop performance to 10-Mbps, and to support high-bandwidth applications in the new superserver-based workgroup.

The solution (see Figure 2.17) migrates existing users from 10-Mbps hubs to switched dedicated 10-Mbps connection to the desktop. (The hubs can easily be redeployed to accommodate growth in other parts of the network.) A fast Ethernet hub provides shared 100-Mbps to the desktop for the new power workgroup. A fast Ethernet switch aggregates the 100-Mbps links and provides a dedicated 100-Mbps connection for the superserver.

Switched Token Ring Technology

Ethernet users have seen a gamut of new technologies being introduced, as discussed throughout this chapter. Token Ring users have seen much less activity. However, two years behind Ethernet switches, Token Ring switches are now being introduced. (It is interesting to note that in spite of the fact that IBM has recently been putting emphasis on its 25-Mbps ATM, Token Ring users will initially experience exposure to LAN switching via Token Ring, and from vendors other than IBM. See Table 2.5 for a listing of typical switch features.) Token Ring switching costs an average of $1500 per port at press time (for comparison, Ethernet switching can be accomplished for $600 to $750 per port).

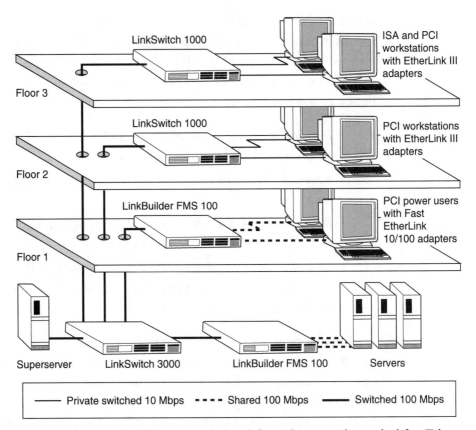

Figure 2.17 *Switched Ethernet and shared fast Ethernet with switched fast Ethernet backbone.*

Token Ring switching is based on Layer 2 switching. As for competitive switching technologies, including ATM, the goal is to increase the available desktop and server-level bandwidth. Token Ring switching can increase the LAN bandwidth per connection without requiring new workstation adapters, new wiring, or changes to existing workgroups or backbones. However, there are limitations associated with Token Ring switching compared to an ATM solution. Specifically, ATM has been designed to support voice, video, and data. While Token Ring switching will increase the throughput for data applications, and while it may be marginally effective for real-time digitized business video, to date it has not supported voice or entertainment-quality video applications. Furthermore, even for data, Token Ring can increase the bandwidth per connection, but not nearly as much as ATM: The maximum bandwidth that can be provided is 32 Mbps, and this requires full-duplex adapters. Hence, as for switched Ethernet, Token Ring switching can be utilized as an interim solution to increase campus-level bandwidth as part of a migration to ATM at the desktop and in the workgroup [8].

Deployment Approaches

There are three typical implementation scenarios for Token Ring switching: (1) standalone workgroup use, (2) segmentation of large LANs, and (3) backbone use.

> *Standalone workgroup use.* LAN workgroup users currently connected to a Token Ring multistation access unit (MAU) or hub can be upgraded by replacing the MAU or hub with a Token Ring switch. With this upgrade, PCs, workstations, or servers are terminated on ports on the Token Ring switch via existing, but now dedicated, STP or UTP wiring. This affords a bandwidth of 4 or 16 Mbps to each LAN device, depending on the speed of the Token Ring NIC. If full-duplex NICs are available, the devices can operate at 32 Mbps: A device (e.g., a server) can receive data from a user attached to the switch at 16 Mbps while simultaneously transmitting data at 16 Mbps to another device.
>
> *Segmentation of large LANs.* In this implementation, cascaded LAN workgroup devices and servers are segmented into smaller LAN segments terminating on a MAU or hub; in turn, the MAUs or hubs are connected to individual ports on the switch. In this configuration, LAN throughout is improved at the segment level (each LAN seg-

Table 2.5 Features of Token Ring Switches

Ability to interconnect multiple switches
Ability to interconnect the switch to a high-speed backbone
Advanced filtering and security to minimize broadcast traffic forwarded by the
   switch
Support of cut-through switching to forward PDUs between ports
Support of a rich set of network management capabilities (e.g., RMON)
Support of SRB

---

ment can have a dedicated 16 Mbps), rather than the individual device level. However, since there will be significantly fewer devices on each of the newly created LAN segments, each device can secure a token and transmit across the LAN much more frequently. Another application of this design approach is to upgrade LAN servers with full-duplex NICs and connect them to dedicated ports on the switch.

Backbone use.   Here, SRBs used to internet distinct Token Ring LANs to the campus backbone are replaced by a Token Ring switch. Switches have faster processors compared to SRBs and, therefore, can forward packets to and from a backbone faster. Also, by upgrading from half-duplex to full-duplex, transmissions occur more rapidly.

Nearly all of the Token Ring switches either available or announced can be used in any of the preceding three scenarios. However, *backbone* and *segmentation* scenarios are the more likely deployment opportunities in the immediate future. Most Token Ring LANs supporting transaction processing traffic (which is a common application for these networks), are not transmission-speed bound. Hence, there is no economic incentive for using switching on individual LAN workgroups, particularly given the relatively high cost of Token Ring switching. Just as with other technologies, Token Ring switches may need to be interconnected either to access more ports than are available on a single switch or to achieve interworking between dispersed LANs. At this early time FDDI, rather than ATM, is being advocated by vendors as the backbone technology.

Switch Features

There are two methods of forwarding PDUs in switched Token Ring environments, just as for switched Ethernet: *cut-through switching* and *store-and-forward switching*. Cut-through PDU forwarding reduces the time required for a switch to forward a complete PDU from one port to another, because the switch starts to forward the bits of the PDU as soon as it determines the destination address and the required output port. In cut-through switching, the latency equates to the time needed to determine the output port and to the processing time to transfer bits between ports across the switch. Latency is typically 30 to 50 ms. However, cut-through switching can forward PDUs that contain errors, which results in inefficient bandwidth utilization. Some switches now use *adaptive cut-through*. Here a PDU error-rate threshold is associated with each port. Should the threshold be exceeded, the switch automatically starts using store-and-forward until the error rate on that port falls below the threshold [8]. In store-and-forward switching, the switch waits until it has received a complete error-free PDU (via frame check sequence calculation) before it begins to forward it to the output port. Latency is typically around 100 ms.

SRB is a common value-added feature occasionally provided on Token Ring switches. The capability can usually be activated on a per-port basis. SRBs enhance the flexibility and overall cost-effectiveness of switching solutions. (As an alternative, the manager seeking to support redundant paths would have to use external SRBs, increasing the cost.) However, SRB features should be used judiciously, since processing SRB-related RIFs increases switch latency.

Most switch vendors provide filtering and address/name caching to minimize the amount of broadcast traffic that propagates between ports. Since these switches operate at Layer 2, the filtering is restricted to Layer 2 objects, such as MAC addresses. Often, such as in TCP/IP and SPX/IPX environments, more robust filtering may be required, covering both the data link layer and the network layer (e.g., for firewalling purposes). An external router may be required.[30] (A small number of vendors include this function in the switch.)

References

1. D. Minoli. *1st, 2nd, and Next Generation LANs.* New York: McGraw-Hill, 1994.

2. D. Minoli and A. Alles. *LANs, ATM, and LAN Emulation.* Norwood, Mass.: Artech House, 1997.

3. IEEE Standard 802.3u. *MAC Parameters, Physical Layer, Medium Attachment Units and Repeaters for 100 Mbps Operation,* v. 4 (supplement to ANSI/IEEE Std. 802.3-1993). IEEE, January 1995.

---

[30] We agree with the following observation: "Whether a Token Ring switch should provide Layer 3 routing functions is debatable, and also begs the question of whether a switch with routing is a bona fide switch or just a 'fast router' masquerading as a switch" [8].

4. D. Trowbridge. "Fast Ethernet Integrators Facing Migration Muddle." *Computer Technology Review* (June 1995): 1–3.

5. A. Chiang. "A Look at 100Base-T." *Communication System Design* (August 1995): 43–45.

6. IEEE Standard 802.12. *IEEE Draft Standard for Information Technology Local and Metropolitan Area Networks—Part 12: Demand-Priority Access Method, Physical Layer and Repeater Specifications for 100 Mbps Operation.* IEEE, March 1995.

7. R. Perlman. *Interconnections: Bridges and Routers.* Reading, Mass.: Addison-Wesley (1992).

8. A. Guruge. "Token-Ring Switches Come of Age." *Business Communication Review* (August 1995): 38–42.

3

ATM and ATM Switching

This chapter focuses on Asynchronous Transfer Mode (ATM), which is a broadband switched service supporting both LAN and WAN applications. This chapter provides a summary view of the topic. The reader requiring more details can refer to any number of texts, including References [1] to [4].

ATM services are being contemplated by chief technology officers (CTOs) in many organizations as possible capabilities in support of corporate networks as a way to face the current trend for lower information systems (IS) expenditures, while at the same time providing increased functionality. In particular, there is an increased need to connect all sites of an organization by a network of sufficient capacity to allow the corporation to conduct its business in an effective manner. In addition, there are increased interenterprise connectivity demands (e.g., over the Internet). Furthermore, the logical topology of the interconnected parties needs to be tailorable to changing business needs; for example, in support of virtual corporation concepts.

Until recently, customers' enterprise networks have consisted mostly of a mesh of dedicated point-to-point lines. These dedicated lines have been migrated from analog and 56-kbps digital lines, up to T1/DS1 and even T3/DS3 lines; these lines were traditionally terminated on T1/T3 multiplexers or directly on routers. Disadvantages of this approach include the cost of the mesh network of dedicated lines. ISDN services, Frame Relay, and Cell Relay are *switched* high-speed services, operating at 1.5, 45, and 155 Mbps,[1] that enable the user to eliminate multiple dedicated lines and

---

[1] In the future some of these services (e.g., cell relay), may operate at a higher speed (622 Mbps or higher).

obtain not only more cost-effective connectivity but also increased reach (since more destinations can be connected without having to install and maintain facilities) and improved availability (since the network can afford alternate routing via a large number of available carrier facilities). Figure 3.1 shows the topological advantages of using a switched service such as Cell Relay: Reduced circuit mileage implies reduced costs to the user. In addition, the reduced number of facilities implies simpler network management.

ATM's Emergence

ATM supports the deployment of integrated intranets and enterprise networks that allow corporations to get their voice, video, and data across town as easily as across the hall, using one network.

High-throughput applications are now entering the mainstream. Examples include scanning documents and imaging, hypermedia documents, multimedia, and desktop video and videoconferencing. In addition, 100-Mbps LANs (e.g., 100-Mbps Ethernet) are entering the corporation. Unconnected islands of users are no longer acceptable across either geographical or technological barriers. Hence, high-throughput WAN and interworking systems are needed.

In particular, ATM platforms and cell relay services are now being considered as the vehicles to support a variety of such evolving applications as follow:

- LAN interconnection over a WAN (bridged) link, particularly for high-speed 100-Mbps LANs
- Legacy-to-broadband interworking via LANE[2]

---

[2] LAN emulation entails emulating the MAC layer functionality of specific LAN protocols, such as Ethernet or Token Ring. This allows end stations on traditional LANs to communicate with end stations on ATM LANs without requiring any hardware or software changes to the end stations on the traditional LANs. But this requires separate MAC emulators for each of the previously mentioned LAN protocols, to be deployed in end stations connected to ATM LANs and in bridges connecting ATM LANs with traditional LANs.

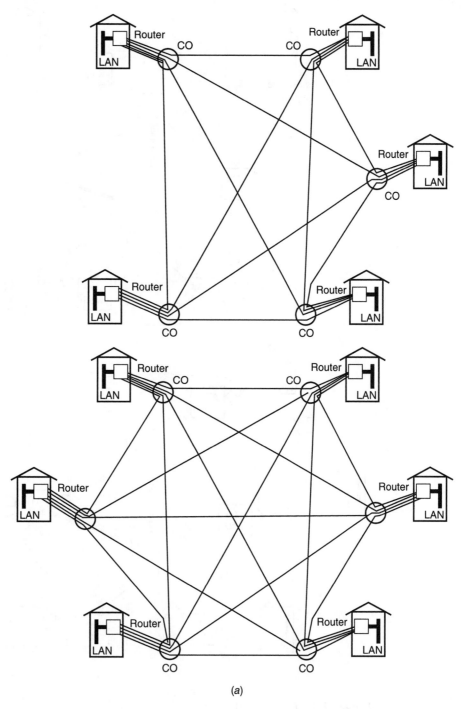

(a)

Figure 3.1(a) Example of enterprise networks comprised of dedicated lines, showing the difficulty of adding a new site to the network. (b) Example of enterprise networks utilizing public switched services.

151

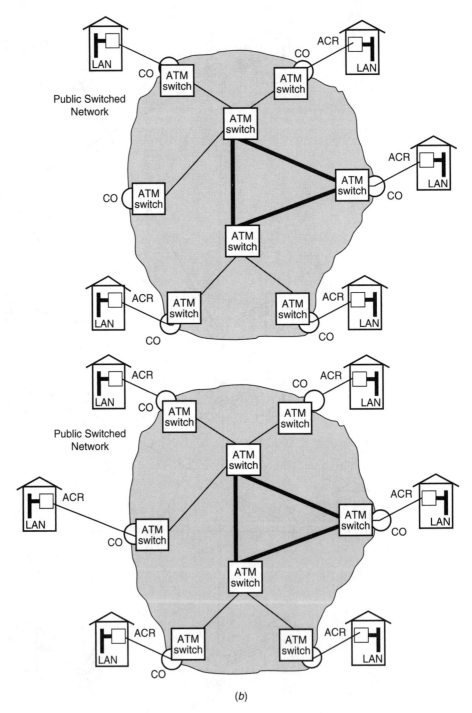

(b)

Figure 3.1 (Continued)

- Multimedia conferencing, collaborative computing, and similar applications
- Image-based workflow systems
- On-demand video distribution using Motion Pictures Expert Group Standard 2 (MPEG-2) over ATM
- Carriage of transport and network protocols (such as Transmission Control Protocol/Internet Protocol [TCP/IP]) over ATM
- Circuit emulation (the realization of a service similar to private lines, without having to put the physical facility in place end-to-end)
- Support of intranets and Internet high-capacity backbones

Routed internets at DS1 speeds or less are running out of steam. DS1 private lines do not scale well to DS3, due to the bandwidth granularities. Also, DS3 does not scale seamlessly to OC-3. Figure 3.2 depicts the technology evolution over the years, in terms of the ubiquitous permeation of complex technologies throughout the corporation.

As discussed elsewhere, LAN segmentation has been used at the LAN level to increase bandwidth per user. However, unless other elements of the enterprise network are appropriately upgraded, the end-to-end throughput will be compromised. L2Ss used in LAN segmentation are fast, but if the routers, building or campus backbone, or WAN backbone are inadequate, the actual end-to-end QoS and throughput will suffer.

ATM uses cellularization in order to make the most effective use of available network bandwidth. Because the cells are small, it is easy to get the maximum utilization of transmission resources. Long PDUs could

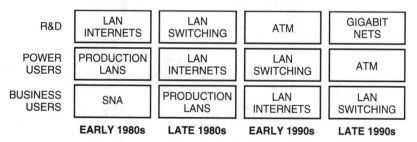

Figure 3.2 Evolution of LAN technology.

imply that there are unfilled gaps between transmissions. Short standardized packets (cells) allow the multiplexing equipment at the link's endpoints to fully pack the link. Figure 3.3 diagramatically represents this. Furthermore, cells of constant length improve the delay performance through the switch, as illustrated in Figure 3.4.

ATM took shape through the full international consultative procedure as well through industry initiatives (e.g., ATM Forum), although the standard specifications do not yet resolve every possible issue related to WAN or LAN networking. This process took about ten years (1985 to 1995), and is ongoing. It grew out of other well-established standards. Researchers in the 1980s were looking at the best method to achieve high-speed packetized data transmission. It began with ISDN standards activities. ISDN is a digital service, but it is characteristic of its generation.

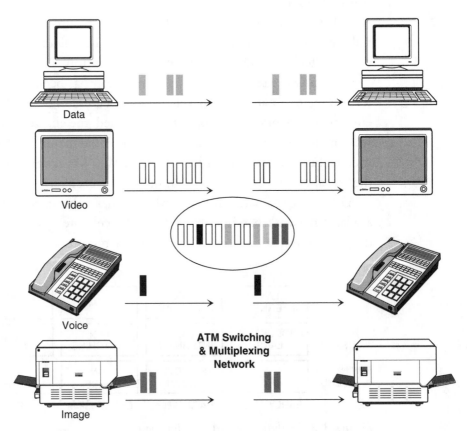

Figure 3.3 *ATM's multimedia potential.*

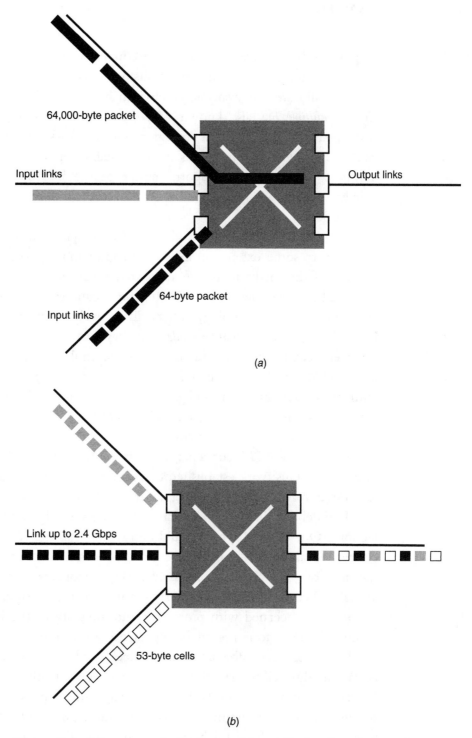

Figure 3.4 *Advantages of cell relaying: (a) variable-length packets, and (b) fixed-length cells.*

Its physical layer (Layer 1) is based on copper wire, and its bandwidth is limited to hundreds of thousands of bits per second.

ISDN allocates the available bandwidth into three channels using *time division multiplexing* (also known as *synchronous transfer mode*) technology: two 64-kbps channels for data transport and one 16-kbps channel for signaling. The signaling channel is used to establish a connection; the data transport channels then support transfer of the information. ISDN is strictly a circuit-switched service of defined bandwidth capabilities ($n \times 64$, $1 \le n \le 30$).

A decision was made in the mid-1980s to seek a new standard that could be based, to some extent, on ISDN principles and support optical fiber; because of the media used, the supported speeds are much higher. The standard became known as *B-ISDN* (for broadband ISDN). Although both are digital, ATM technology differs distinctively from ISDN insofar as ISDN is a *synchronous transfer mode technology*—namely, ISDN is a circuit-switched technology without any statistical multiplexing and statistical gains. ATM, on the other hand, is an *asynchronous transfer packet technology* with statistical multiplexing and gain. Because the user and the carrier gamble on statistical multiplexing, sophisticated traffic management capabilities are required on the network switches.

Fundamental ATM concepts arose from research conducted in the mid-1980s. This work established that data units of fixed length are easier to switch at very high speeds than frames that can vary in length. The B-ISDN debates that led to standardization were carried on between two factions. One faction, which could be identified as the *X.25 faction,* was concerned above all with more efficient, versatile, and less costly data transmission. The *data faction* was relatively unconcerned about time of arrival and could tolerate some delay while packets are reassembled. They were very concerned with protecting the integrity of the information. The public telephone providers, representing the other faction, emphasized time-sensitive information like voice, in which consistent sequence is crucial—limited information loss was acceptable to them. There were, however, common grounds. For packetized video, packets must come through at a regular rate in order to avoid transmission *jitter.* This is easier to accomplish with relatively compact packets. The two factions solved this problem by deciding on a fixed-length packet, called a *cell,* that could be transmitted in an orderly, high-speed fashion over a switched network. This solution would provide the cost advantages of data networks combined with the predictability of voice networks. But how long should those cells be? The data faction advocated a 64-byte specification, while the voice faction demanded 32 bytes. An agreement was finally reached wherein each cell would contain a 48-byte payload, accompanied by 5 additional bytes to identify the cell and carry other protocol control information.

Several entities have published relevant ATM standards, specifications, and requirements, including International Telecommunication Union—Telecommunications (ITU-T) and ANSI (ATIS). The ATM Forum (ATMF), the Frame Relay Forum, and Bellcore have published relevant documentation. Carriers require the support of the ATMF User-to-Network Interface Specification Version 3.1 or 4.0; Broadband Inter-carrier Interface (B-ICI) Version 2; and (for internal connectivity only) Private Network Node Interface (P-NNI) Version 1.0 or higher. UNI 3.1 supports the ITU-T Q.2931, which is important. LAN Emulation Version 1.0 or 2.0 and MPOA standards are also required for turn-of-the-century networks.

For high-bandwidth networks based on fiber, ATM is frequently employed with a Layer 1 standard, called *SONET*.[3] SONET defines a series of bandwidth levels for transmission over fiber networks. SONET rates consist of multiples of a base of 51.840 Mbps. Current ATM technology supports bandwidths at the Optical Carrier 1 (OC-1) level (51.840 Mbps), OC-3 level (155.520 Mbps), and at the OC-12 level (622.08 Mbps). SONET levels now targeted by systems developers would deliver 1244.160 Mbps (OC-24) and 2488.320 Mbps (OC-48). SONET standards are now nearly ready for OC-192 (about 10 Gbps) speeds. For user access, current ATM technology supports only 155.520 Mbps; the 622.080 Mbps speed is currently supported only at the trunk level. OC-24 and OC-48 would represent the gigabit network concept mentioned so often in public discussion

---

[3] Outside the United States, the SONET concept is described in the context of the Synchronous Digital Hierachy (SDH); effectively, this hierarchy uses building blocks of 155.520 Mbps rather than building blocks of 51.840. However, they are basically consistent for appropriate values of the aggregate bandwidth.

of information superhighways. Applied with ATM technology, these band-width levels should be achievable in this decade.

As discussed, accredited standards bodies have developed the basic set of protocols. As is always the case, implementer's workshops are then needed to complete the implementation details. In the case of ATM, the ATM Forum has focused on issues of interoperability. It appears that ATM will resolve these vendor interoperability issues more quickly than was the case for its predecessors. This is primarily due to two factors: focus and com-mitment. The ATM Forum consists of a plethora of vendor, user, govern-ment, and academic representatives whose commitment to the success of ATM is, in some people's view, unparalleled. FDDI, for example, has encountered some early difficulties in linking workstations from different manufacturers. Token Ring also had problems of this kind. This was due to the lack of an organization focusing efforts on implementation and deployment issues.

Because it can utilize fiber media to its full potential (at least in the long haul) ATM supports high speed. In the campus environment, ATM uses multimode fiber to derive 100 Mbps (using FDDI-like encoding) or single-mode fiber to derive 155 Mbps (using FC-like encoding), thereby not utilizing the fiber to its full potential; it also uses UTP 5 twisted-pair to sup-port speeds up to 155 Mbps. Hence, ATM provides high speed when and only when it uses the full potential of the underlying media. These larger transmission capacities, used in conjunction with appropriate switching technology, make it easier to cope with bursty traffic. Ethernet, with its 10-Mbps ceiling, can also handle bursty traffic but is hindered by the con-tention issue and the easy-to-reach ceiling—there are transmission ceilings (now 622 Mbps) and switching ceilings (now 2 to 20 Gbps) in ATM, but these are, at this time, more difficult to overwhelm. Token Ring and FDDI reduce contention compared to Ethernet, but have much less capability to handle bursty transmissions. ATM's ability to allocate bandwidth dynami-cally makes it easier to offer different classes of service to support different application classes. The transparency of the upper layers to ATM, when appropriate adaptation is provided in the pertinent equipment, makes it possible for ATM to be ubiquitous, employed in multiple types of LAN, backbone campus, and WAN topologies.

Fiber media, used with the SONET Physical Layer protocol, makes high capacity available to the connected users. However, there is still a need to support statistical multiplexing and bandwidth overbooking in order to obtain transmission efficiencies, particularly in the long haul. It follows that in ATM sophisticated congestion control is needed in order to guarantee a very small probability of cell loss, even under significant traffic levels. These characteristics add up to a technology that can permit users and applications developers to explore possibilities that have not been feasible until now.

Overview of Key ATM Features

This section outlines some of the principal features of ATM. The features discussed here are as follows:

- The structure of its 53-byte cells, or labeled information containers.
- The Physical Layer, ATM Layer, and ATM Adaptation Layer (AAL) that organize appropriate service data units (SDUs) and PDUs for transmission. Special attention is given to the Adaptation Layer, which is used to accommodate the special requirements of voice, video, and data traffic. The service layer sits on top of the AAL and uses specific AALs (e.g., AAL 1, AAL 5, etc.) to provide the appropriate services to the legacy protocols (e.g., IP) residing at the network layer.
- LAN emulation, in support of legacy LANs.

The ITU-T standards for ATM specify the cell size and structure and the UNI. Note that there are two kinds of UNIs: one for access to public networks, and one, called *Private UNI,* for access to a customer-owned ATM network (specifically to a hub, router, or switch). As already discussed, for the public UNI the physical layer is defined for datarates of 1.544, 45, and 155 Mbps (SONET OC-3). For Private UNI, a number of physical layers for different media (UTP, STP, single-mode fiber, and multimode fiber) are defined.

ATM can be described as a packet transfer mode based on asynchronous time division multiplexing, and a protocol engine that uses small, fixed-length data units—namely, cells. ATM provides a connection-oriented service. Note that LANs such as Ethernet, FDDI, and Token Ring support a connectionless service. Hence, interworking must take this into account. Each ATM connection is assigned its own set of transmission resources; however, these resources have to be taken out of a shared pool that is generally smaller than the maximum needed to support the entire population—this is the reason for the much-talked-about traffic management problem in ATM. ATM nevertheless makes it possible to share bandwidth through multiplexing (multiple messages transmitted over the same physical circuit). Multiple virtual channels can be supported on the access link, and the aggregate bandwidth of these channels can be overbooked (ATM relies on statistical multiplexing to carry the load). Within the network, expensive resources are rationed, and bandwidth must be allocated dynamically. ATM is thus able to maximize resource (bandwidth) utilization.

ATM supports two kinds of channels in the network: *Virtual Channels* (VCs) and *Virtual Paths* (VPs). VCs are communication channels of specified service capabilities between two intermediary ATM peers. *Virtual Channel Connections* (VCCs) are concatenations of VCs to support end system–to–end system communication. VPs are groups (*bundles*) of VCs. *Virtual Path Connections* (VPCs) are concatenations of VPs to support end system–to–end system communication. In VC switching, each VC is switched and routed independently and separately. VP switching allows a group of VCs to be switched and routed as a single entity. This concept only applies to ATM and not to the other services available over the ATM platform.

A virtual circuit can be either switched (temporary) or permanent. A connection is established through preprovisioning with the carrier or private devices (thereby establishing *Permanent Virtual Channels* [PVCs]), or through signaling mechanisms (thereby establishing *Switched Virtual Channels* [SVCs]). Connections supported by these channels (PVCs or SVCs) enable one computer or other system on the network to communicate with another. More specifically, communication in ATM occurs over a concatenation of virtual data links (VCs)—this comprises a VCC. VCCs can be permanently established by an external provisioning process, entailing a service order (with desired traffic contract information) and manual switch configuration. Such a VCC comprises a PVC. Evolving switched cell relay service (based on ATM principles) requires signaling to specify the details of the connection, such as the forward and backward bandwidth, the quality of service class, the type of adaptation and interworking, subaddressing, point-to-multipoint connectivity, and other user-to-user information. When the control plane mechanisms are implemented in both the user equipment and in the switch (specifically ITU-T Q.2931 and ATMF 3.1), the user will be able to establish connections automatically on an as-needed basis. This comprises SVC connection.

Figure 3.5 depicts a functional reference model that is utilized when developing recommendations and deploying equipment.

Network resources such as inbound speed, outbound speed, quality of service, multipoint capabilities, and so on are requested as a connection is established. A connection is established if the network is able to meet the request; if not, the request is rejected. Once the virtual circuit is defined, the call connection control assigns an interface-specific *Virtual Channel Identifier* (VCI) and *Virtual Path Identifier* (VPI) to identify the connection. These labels have only interface-specific meaning. Two different sets of

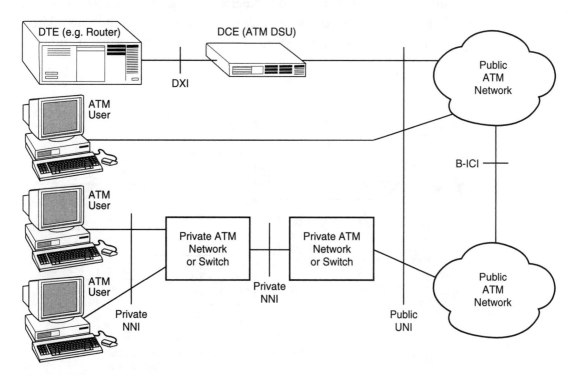

Figure 3.5 *Functional reference model.*

VPIs and VCIs are assigned to the two endpoints of the connection. Inside the network, as many sets of VPIs and VCIs as needed along the path are used by the network, but are invisible to the end users. As long as the connection remains active, the assigned VCI and VPI represent valid pointers into routing tables in the network; the tables (accessed via the VPI and VCI) are used to accomplish cell routing through the network. Figure 3.6 depicts a simple example of cell switching using VPI/VCI lookup.

SVCs are utilized to support native or interworking services. Specifically, LANE is an interworking capability that allows Ethernet and Token Ring stations to communicate directly with ATM stations (and vice versa) as if they were using the same protocol. The interworking equipment supports the conversion between the two protocols. Traditional LANs use the 48-bit MAC address. The MAC address is globally unique. This nonhierarchical LAN address, assigned by the manufacturer, identifies a network interface in the end station. The use of a MAC address is practical in a single LAN segment or in a small internet. However, large bridged networks become difficult to manage and experience large amounts of broadcast traffic for the purpose of attempting to locate end stations. The address space of a large network is preferably hierarchical. This makes it easier to

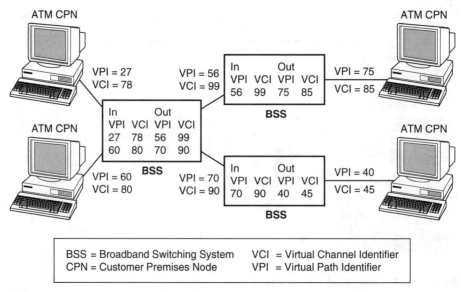

Figure 3.6 *Switching ATM Virtual Connections.*

locate a particular point on the network; such an address, however, restricts the mobility of the network users. The E.164 address used in public ATM is hierarchical.

To emulate a LAN, the ATM network must support addressing using the MAC address scheme: Each ATM MAC entity must be assigned a 48-bit MAC address, from the same address space, to facilitate its identification. As noted, an ATM network, whether public or private, uses a hierarchical address. The address resolution operation in LANE binds the end station MAC address to the physical address of the ATM port to which the end station is currently connected. When an end station is attached to an ATM switch port, a registration protocol exchanges the MAC address between the ATM network and the end station. The LAN emulation service consists of several pieces of software and hardware operating on one or more platforms, as follows:

- *LAN Emulation Client (LEC).* The LEC is software that resides at the edge device. The *edge device* is where the emulated service is rendered in terms of the conversion between protocols.
- *LAN Emulation Server (LES).* The LES provides initialization and configuration functions, address registration, and address resolution. Since ATM and legacy LANs use very different addressing schemes, a way to map the two is important, particularly with a view to *subnetworks,* where the addressing capabilities of ATM may be lacking.
- *Broadcast and Unknown Server (BUS).* The BUS provides the mechanism to send broadcasts and multicasts to all devices within the emulated LAN.

In a traditional LAN, all frames (unicast, multicast, and broadcast) are broadcast to all stations on the shared physical medium; each station selects the frames it wants to receive. A LAN segment can be emulated by connecting a set of stations on the ATM network via an ATM multicast virtual connection. The multicast virtual connection emulates the broadcast physical medium of the LAN. This connection becomes the broadcast channel of the ATM LAN segment. With this capability, any station may broadcast to all others on the ATM LAN segment by transmitting on the shared ATM multicast virtual connection.

Cell Format

The ATM cell has a 48-byte payload, accompanied by a 5-byte header that is divided into fields. Headers are of two types: the UNI and the Network-to-Network Interface (NNI). See Figure 3.7.

Fields within the UNI cell are as follows:

- The first field, of 4 bits, provides for *generic flow control* (GFC). It is not currently used and is intended to support a local bus (extension) function to connect multiple broadband terminal equipment to the same UNI as equal peers. (Note that multiple users can be connected to the UNI today by using a multiplexing—not peer—function.) This is equivalent to the SAPI function in ISDN.

- A 24-bit routing pointing field is subdivided into an 8-bit *VPI* subfield and a 16-bit *VCI* subfield. It indirectly identifies the specific route laid out for traffic over a specific connection, by providing a pointing function into switch tables that contain the actual route.

- Three bits are allocated to the *payload type identifier* (PTI), which identifies whether each cell is a user cell or a control cell, used for network management.

- A single-bit *cell loss priority* (CLP) marker is used to distinguish two levels of cell loss priority. Zero identifies a higher priority cell that should receive preferred loss treatment if cells are discarded due to

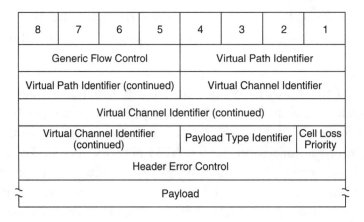

Figure 3.7 *ATM cell structure.*

network congestion. One indicates lower-priority cells whose loss is less critical.

- *Header error control* (HEC) is an 8-bit cyclic redundancy code (CRC) computed over the ATM cell header. The HEC is capable of detecting all single-bit errors and certain multiple-bit errors. It can be used to correct all single-bit errors, but this is not mandatory. This mechanism is employed by a receiving device to infer that the cell is in error and should simply be discarded. It is also used for cell-boundary recovery at the physical layer.

- The remaining 48 bytes are devoted to *payload*. (In case of interworking, some of the payload bytes have to be used for the network's AAL protocol control information.)

- The NNI cell structure has one difference: The 4-bit GFC field is dropped, and the VPI field is expanded from 8 to 12 bits.

Addressing

Addressing is a fundamental need in any network. The ITU-T ATM protocol calls for a hierarchical ISDN telephone numbering scheme, specified in ITU-T E.164, to be used in ATM. The standard permits the ATM address to be divided into an address and a subaddress. The ATM Forum recommends that the address describe the point of attachment to the public network (if connected to the public network) and that the subaddress identify a particular end station within a private network [5]. Note that the VPI and VCI are just labels, not E.164 addresses—they are table pointers for relaying cells on to their destinations, based on switch-routing tables. The ATM Forum specification permits two address formats to be used as ATM addresses in private networks. One is the E.164 format, and the other is a 20-byte address modeled after the address format of an OSI Network Service Access Point (NSAP).

The Protocol Model

An extension of the conventional OSI seven-layer stack can be used to describe the structure of the ATM protocol; a reference model specific to

ATM depicts its structure more clearly. Figure 3.8 depicts the B-ISDN protocol model, while Figure 3.9 depicts the protocol stack used in a simple ATM network. This reference model distinguishes three basic layers. Beginning from the bottom, these are the *Physical Layer,* the *ATM Layer,* and the *ATM Adaptation Layer* (AAL). Each is divided further into sublayers, as seen in Figure 3.10. To further elaborate the protocol stack, Figure 3.11 shows both the data flow plane and the signaling plane required to support switched connections.

The Physical Layer includes two sublayers:

- Like any other data link layer protocol, ATM is not defined in terms of a specific type of physical carrying medium, but it is necessary to define appropriate physical layer protocols for cell transmission. The *physical medium* (PM) sublayer interfaces with the physical medium and provides transmission and reception of bits over the physical facility. It also provides the physical medium with proper bit timing and line coding. There will be different manifestations of this layer based on the specifics of the underlying medium (e.g., DS1 link, DS3 link, SONET, UTP, etc.)

- The *transmission convergence* (TC) sublayer receives a bit stream from the PM sublayer and passes it in cell form to the ATM layer. Its

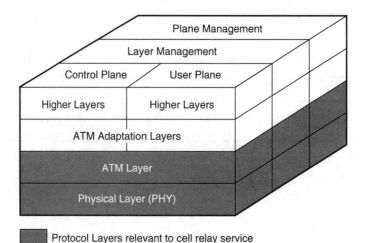

Protocol Layers relevant to cell relay service

Figure 3.8 *B-ISDN Protocol reference model.*

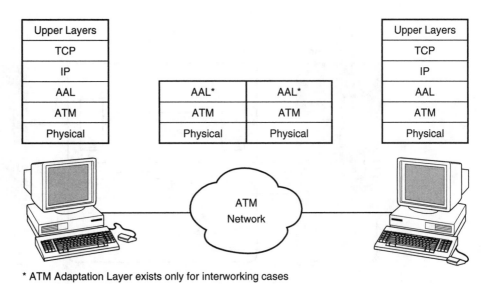

* ATM Adaptation Layer exists only for interworking cases

Figure 3.9 *ATM peers.*

Convergence	CS	AAL
Segmentation and Reassembly	SAR	
Generic Flow Control (if and when implemented) Cell VPI and VCI Translation Cell Multiplex and Demultiplex		PHY
Cell Rate Decoupling HEC Header Sequence Generation and Verification Cell Delineation Transmission Frame Adaptation Transmission Frame Generation and Recovery	TC	PHY
Bit Timing Physical Medium	PM	

Figure 3.10 *Signaling stack.*

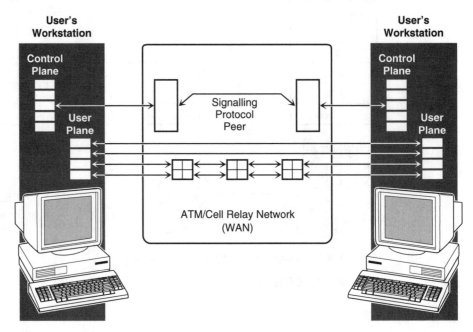

Figure 3.11 *ATM's dual stack.*

functions include cell rate decoupling, cell delineation, generation and verification of the HEC sequence, transmission frame adaptation, and the generation and recovery of transmission frames.

The ATM layer, in the middle of the ATM stack, is responsible for one of ATM's most "trivial" functions—to encapsulate downward-coming data into cells from a number of sources and multiplex the cell stream. Conversely, it has the responsibility to de-encapsulate upward-coming cells and demultiplex the resulting stream out to a number of sources.

The ATM layer controls multiplexing (the transmission of cells belonging to different connections over a single cell stream) and demultiplexing (distinguishing cells of various connections as they are pulled off the flow of cells.) ATM, as a data link layer protocol, is medium-independent: It is capable of performing these functions on a wide variety of physical media. In addition, the ATM layer acts as an intermediary between the layer above it and the physical layer below. It generates cell headers, attaches them to the data delivered to it by the adaptation layer, and then delivers the properly tagged cells to the physical layer. Conversely, it strips headers from cells containing data arriving on the physical layer before hoisting the data to the application layer.

VCs and VPs are identified by their VCI or VPI tags.[4] The ATM layer assures that cells are arranged in the proper sequence, but it does not identify and retransmit damaged cells. If this is to be done, it must be accomplished by higher-level procedures. The ATM layer also translates VCI and VPI information. As noted earlier, each ATM switch has its own routing table to identify each connection. In transit between switches, VPI and VCI identifiers (routing table pointers) will be different. Switches translate identifiers as they transfer cells onward to other switches.

Finally, the ATM layer performs management functions. If the PTI identifies a cell as a control packet, the ATM layer responds by carrying out the appropriate functions.

Support of Existing Protocols over ATM

The ATM Adaptation Layer allows various network layer protocols to utilize the service of the ATM layer. As discussed earlier, the ATM layer only supports the lower portion of the data link layer. Hence, in order for the network layer to use ATM, a *filler sublayer* is required. This is analogous to IP use over a LAN: The Media Access Control layer only supports the lower portion of the data link layer; consequently, the Logical Link Control layer is sandwiched in between.

Fundamentally, the AAL keeps the network layer "happy" by enabling it to use ATM transparently. The basic function of the AAL is to segment the downward-coming data (network layer PDU) into cells, and to reassemble upward-coming data into a PDU acceptable to the network layer. It is critical to understand that AALs are end-to-end functions (end system–to–end system). A network providing pure ATM will not be aware of or act upon

---

[4] Do not confuse VCs and VPs with connections. VCs are *channels—connections* are instances of end-to-end communications. Connections are identified by Call Reference and Connection Identifiers included in the Setup message used in signaling. See [6] for a more extensive description.

AAL information (but in the case of service interworking in the network, there has to be network interpretation of the AAL information).

In one classical view of the ATM protocol model, a *service* layer resides above the AAL, in the end systems. Hence, by further elaboration one can say that, in a coincidental manner, the AAL differentiates the treatment of different categories of cells in the end system and permits responses to user-to-user quality-of-service issues. A number of AALs has been defined to meet different user-to-user quality-of-service requirements. Again, however, a network providing pure ATM will not be aware of or act upon AAL information (but in the case of service interworking in the network, there has to be network interpretation of the AAL information). Therefore, the AAL-supported service differentiation is among end system peers, and is not the mechanism used by the ATM network to support network quality of service. We use the term *user-to-user QoS* to describe the kind of end system–to–end system peer-to-peer connection service differentiation. (This connection is viewed as being *external* to the ATM network.)

For example, an end system TV monitor needs a continuous bit stream from a remote codec in order to paint a picture; it may have been decided that an ATM network is to be used to transport the bits. Because of the codec and monitor requirements, the bits have to be enveloped in such a manner that clock information is carried end-to-end in such a fashion that jitter is less than some specified value. To accomplish this the bits are enveloped using AAL 1. From the ATM network's point of view, this is totally immaterial: The ATM network receives cells and carries them to the other end—the network delivers cells. The network does not render any different type of QoS to these cells based solely on the fact that the cells had AAL 1 information in them; the network was not even aware of the content. Naturally, it would be desirable for the network to provide reasonable QoS to this stream, based on some kind of knowledge or arrangement. How is that accomplished?

The different QoS obtained via an ATM network is based on user-to-network negotiation, not by the content of the cell. In PVC this negotiation is via a service order. Here, the user would tell the network (on paper) that the user wanted to get reasonable service for a certain stream carrying codec video. The network provider would make arrangements to terminate this stream on a switch line card where, for example, a lot of buffers are allocated. The network provider then would tell the user (on paper) to employ a certain VPI/VCI combination (say 22/33) for this specific stream. Here is what happens: The user sends cells over the physical interface terminated at the card's port. Certain cells arriving on the interface have VPI/VCI = 44/66; these get some kind of QoS treatment. Then some cells arrive on the interface with VPI/VCI = 22/33; these cells get the agreed-to QoS by receiving specific treatment by the switch. In SVC, a similar mechanism is in place, except that instead of communicating the information on paper, the call-setup message is used (with automatic call negotiation).

In any event, the QoS in the ATM network is not based on the fact that the cells carried a certain AAL. It is the other way around. The user needs a certain end system–to–end system QoS. The user then needs to do two things: select its network-invisible AAL, and separately inform the carrier of the type of QoS needed.

AALs utilize a small portion of the 48-byte payload field of the ATM cell by inserting additional control bits. In all AALs, the ATM header retains its usual configuration and functions. Note that the data coming down the protocol stack is first treated by the AAL by adding its own header (protocol control information). This AAL PDU must, naturally, fit inside the ATM PDU. Hence, the AAL header must fit inside the payoad of the lower layer—here, ATM. To say that quality-of-service definitions are obtained at the cost of reductions in payload is not exactly correct: AALs provide an appropriate segmentation and reassembly function—QoS is supported by the network switch—and, as discussed, AAL classes support peer-to-peer connection differentiation.

In some instances, users determine that the ATM layer service is sufficient for their requirements, so the AAL protocol remains empty. This occurs, for example, if the network layer protocol can ride directly on ATM (this is unlikely for legacy protocols), or if the two end systems do not need additional coordination. In the majority of cases, however, this AAL layer is crucial to the end system protocol stack because it enables ATM to accommodate the requirements of voice, image or video, and data traffic, while providing different classes of service to meet the distinctive requirements of each type of traffic.

Two sublayers make up the AAL: the *segmentation and reassembly sublayer* (SAR) and the *convergence sublayer* (CS). The SAR sublayer segments higher-layer information into a size suitable for cell payloads through a virtual connection. It also reassembles the contents of cells in a virtual connection into data units that can be delivered to higher layers. Functions like message identification and time/clock recovery are performed by the CS sublayer.

AAL 5 is specifically designed to offer a service for data communication with lower overhead and better error detection. It was developed because computer vendors realized that AAL 3/4 was not suited to their needs. In addition to the header, AAL 3/4 takes an additional 4 bytes for control information from the payload field, reducing its capacity by 8.4 percent. They also maintain that the error detection method of AAL 3/4 does not cope adequately with issues of lost or corrupted cells. See Table 3.1.

With AAL 5, the CS sublayer creates a *CS protocol data unit* (CS-PDU) when it receives a packet from the higher application layer. The first field is the *CS information payload* field, containing user data. The *PAD* field assures that the CS-PDU is 48-bytes aligned. A 1-byte *control* field remains undefined, reserved for further use. The 2-byte *length* field indicates the length of information payload, and the *CRC* field is used to detect errors.

When the CS sublayer passes the CS-PDU to the SAR sublayer, it is divided into many *SAR protocol data units* (SAR-PDUs). The SAR sublayer then passes SAR-PDUs to the ATM layer, which carries out transmission of the cell.

When passing on the final SAR-PDU within the CS-PDU, SAR indicates the end of the CS-PDU transfer by setting the payload type identifier (PTI) in the header to 1. By using the CS length field and the cyclic loss redundancy code (CRC) in the header error control (HEC), the AAL can detect the loss or corruption of cells.

Figure 3.12 depicts in simplified form the functionality needed to support legacy systems over ATM. As seen in this figure, AALs are utilized to

Table 3.1 AAL Type 5 CS-PDU

INFORMATION PAYLOAD	PAD	CONTROL	LENGTH	CRC-32
0–64K	0–47 bytes	1 byte	2 bytes	4 bytes

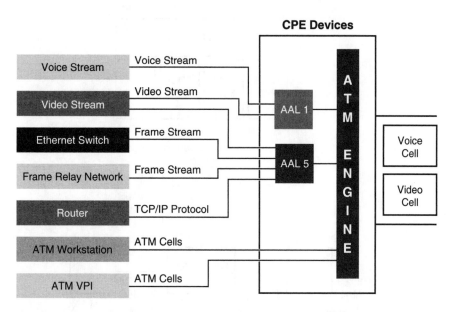

Figure 3.12 *ATM Adaptation Layer services.*

this end. Building on the AAL 5 format, Figure 3.13 shows the encapsulation of an Ethernet frame into cells. Figure 3.14 takes an even more general view of how LAN would work (in LANE 1.0, the encapsulation actually uses VC muxing; in LANE 2.0 SNAP/LLC is utilized). Figure 3.15 shows how AALs can be utilized to support video applications.

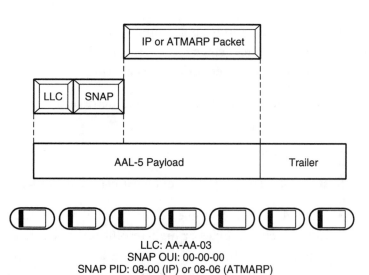

Figure 3.13 *LLC and SNAP encapsulation.*

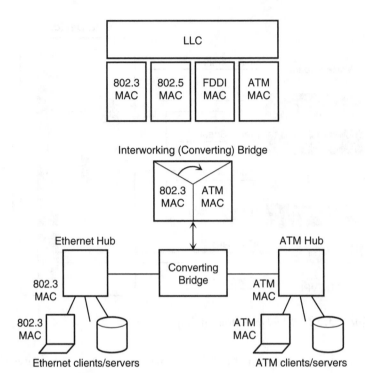

Figure 3.14 *Simplified view of LAN emulation.*

Classes of Service and QoS

The ITU-T specifications apply three broad criteria to distinguish four classes of ATM service, tagged as Classes A, B, C, and D (see Table 3.2). The three end-to-end, network-external criteria are as follows:

- Time relation between source and destination
- Bit rate
- Connection mode

The four end-to-end, network-external classes of service are as follows:

- *Class A.* (For example, clear-channel voice and fixed bit rate video, such as movies or high-resolution teleconferencing.) A time relation exists between the source and the destination. The bit rate is constant, and the network layer level service is connection-oriented.

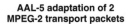

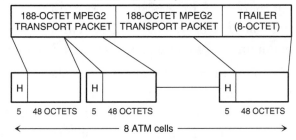

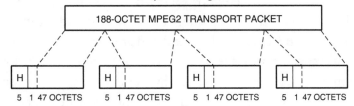

Figure 3.15 *AALs for video over ATM.*

- *Class B.* As in Class A, there is a time relation between the source and the destination, and the network layer level service is connection-oriented, but the bit rate can be varied. Examples include audio and video with variable bit rates (e.g., unbuffered video codecs with motion compensation).

Table 3.2 ATM Classes of Service

	CLASS A	CLASS B	CLASS C	CLASS D
Application	Voice clear channel	Packet video	Data	Data
Timing (source to destination)	Needed	Needed	Not needed	Not needed
Mode	Connection-oriented	Connection-oriented	Connection-oriented	Connectionless
Bit rate	Constant	Variable	Variable	Variable

- *Class C.*   The network layer level service is connection-oriented, but there is no time relation between the source and the destination, and the bit rate is variable. This can, for example, meet the requirements of connection-oriented data transfer and signaling.

- *Class D.*   Intended for applications like connectionless data transport (at the network layer). None of the three parameters applies: Service is connectionless, there is no time relation between the source and the destination, and the bit rate is variable.

These classes are general descriptions of types of user traffic. They do not set specific parameters or establish values. Equipment from multiple vendors based on different parameters may thus find it difficult to establish connections. AALs are end-to-end and are generally external to the ATM network; considerations regarding AAL relate to consideration of end-user equipment. AALs are considered by the network only when there is service interworking. Examples include frame relay–to–ATM interworking in the network, legacy LAN–to–ATM interworking (specifically, LANE) in the network, and private line–to–ATM interworking in the network. For example, in the first case, frames come in and cells go out. There is another case where AALs are used in the network, but this is totally transparent to the user—this situation (sometimes called *network interworking*) is when the network supports a *carriage function* over ATM. Examples include frame relay carriage over an ATM network, Ethernet carriage over an ATM network (e.g., Ethernet bridging), and private line carriage over an ATM network. For example, the frame relay user gives a frame relay frame to the ATM-based network; the network takes the frame and segments it into a stream of cells utilizing AAL 5 protocols; the stream is carried accross the network and to the proximity of the destination; the cells are reassembled into a frame relay frame using AAL 5; and the destination is handed a frame—this type of service is called *frame relay carriage over ATM* or *frame relay–to–ATM network interworking.*

As previously discussed, three AAL protocols have been defined to support the three classes of service in the end system: AAL 1, AAL 3/4, and AAL 5. Computers, routers, and other devices must employ the same AAL in order to communicate with one another on an ATM network.

- *AAL 1* meets the performance requirements of Service Class A. It is intended for voice, video and other constant bit rate traffic, and its performance, to the upper layers of the end system stack, is similar to today's digital private lines. Four bits in the payload are allocated to sequence number (SN) and sequence number protection (SNP) functions.

- *AAL 2* aims at Class B requirements; it has not yet been defined.

- *AAL 3/4* is intended for connectionless data services (e.g., for support of Switched Multimegabit Data Service). Four bytes are devoted to control functions, including a multiplexing identifier as well as segment type (ST) and SNP indicators.

- *AAL 5* is also intended for data communications, including services like frame relay, LANE, and MPOA. The ATM Forum and IETF recommend that AAL 5 also be used to encapsulate IP packets in the user's end system.

Another way of looking at different ATM services is to talk about Continuous Bit Rate (CBR), Variable Bit Rate (VBR), Available Bit Rate (ABR), and Unspecified Bit Rate (UBR) services. These native ATM services support different user requirements, in a fashion somewhat related to the previous service class discussion. *CBR* provides constant bit rate in support of a service that provides the equivalent of a private line at T1 or T3 rates. *VBR* is a variable bit rate service in support of data applications such as frame relay and LAN interconnection. *ABR* provides available bit rate service on a discount basis; here, the amount of bandwidth in the network is not guaranteed, and is managed via a feedback mechanism. *UBR* is unspecified bit rate with only "best-effort" characteristics.

One of the key features of ATM is that it supports negotiated connections. Service is obtained via traffic contract on a per-VC basis. Parameters include traffic characteristics, peak cell rate, and sustainable cell rate. QoS includes delay and cell loss. For ATM services, the customer can specify the equivalent of the sustainable cell rate (SCR), the peak cell rate (PCR), and the burst tolerance (BT). The switch allocates various resources (e.g., trunk capacity and buffers) on a statistical basis. This involves the use of traffic shapers, traffic policers, and tagging for discard. The carrier has the obligation to deploy enough resources in the network to guarantee the type of QoS and the kind of service required (e.g., Constant Bit Rate, Variable Bit Rate, Available Bit Rate, Unspecified Bit Rate, etc.). Non-ATM services are allocated in a similar manner, but the user has no direct control on the ATM traffic parameters.

Related to QoS, one may ask if ATM supports multimedia, and if so, how? Video and multimedia information can be carried over either an ATM UNI, an FDDI interface, or an Ethernet interface. The ATM switch supports a high grade of service, which makes it a good platform for these media.

Also, one can ask if ATM supports multipoint services, and if so, how? ATM supports multipoint connections. The ATFM UNI 3.1 specifies the use of signaling messages to establish such connections. Many ATM switches do not currently support ATM UNI multiconnections (though ATM does support a kind of internal multipoint service for LAN support).

Carrier Infrastructure

A short list of infrastructure considerations follows.

Physically, how are different services supported?   ATM services are supported over a fiber DS3/OC-3 link; the user adds equipment supporting the ATM peer. Frame relay services are supported over a T1 line; the user adds equipment supporting the frame relay peer. LAN-based services are supported by installing the concentrator at the customer's site. The concentrator is connected with the network over an appropriate ATM trunk; the user connects to the concentrator over an RJ-45 and/or SAS/DAS FDDI connection.

What user equipment is needed to access a carrier's cell relay or LAN bridging service?   For ATM cell relay, the user can use an ATM-ready hub, an ATM-ready router, or a workgroup switch. For a legacy service, the user delivers an Ethernet or FDDI connection to the carrier's on-site concentrator.

What is the relevance of P-NNI? Does it matter that our switch does not communicate over the P-NNI to another switch?   Carriers have not, heretofore, supported access by the user to the trunk side of their switches (e.g., voice switches, packet switches, frame relay switches, and SMDS switches). They are not likely to allow access to the trunk side of an ATM switch. Carriers may offer only a UNI service, not a P-NNI service. Hence, it does not matter that a switch may not interwork with another switch over the NNI: This is not relevant to the service a carrier provides. The only relevance of P-NNI is in the event that the carrier decides to use P-NNI as an NNI to connect equipment *within* the carrier's network (but not outside of it).

Traffic Management Version and Congestion Control

An ATM network simultaneously transports a wide variety of network traffic—voice, data, image, and video. It provides users with a guaranteed quality of service (QoS), monitors network traffic congestion, and controls traffic flow to ease congestion. Traffic management allows an ATM network to support traffic flows not readily supported by other LAN technologies. A carefully planned ATM network can guarantee quality of service to each type of traffic as the network grows. This section expands on the previous discussion in terms of describing traffic management and congestion control.[5]

Traffic management in ATM networks consists of those functions that ensure that each connection receives the quality of service it needs and that the flow of information is monitored and controlled within the ATM network. As an organization grows, its business operations become more sophisticated and its information systems requirements must reflect that change. Networks that support a wider variety of traffic types are needed by these organizations. Understanding traffic management in an ATM network requires an appreciation of the differences in the types of traffic it carries.

Classifying ATM Network Traffic

The types of traffic supported by an ATM network can be classified by three characteristics: bandwidth, latency, and cell delay variation. *Band-*

---

[5] This section is based on promotional material from Fore Systems, Warrendale, Pa. Used with permission.

width is the amount of network capacity required to support a connection. *Latency* is the amount of delay associated with a connection. Requesting low latency in the QoS profile, for example, means that the cells need to travel very quickly from one specific point in the network to another. *Cell delay variation* is the range of the delays experienced by each group of cells associated with a given transmission. Requesting low cell delay variation means that the cells in this group must travel through the network without getting too far apart from each other.

As previously discussed, ATM networks carry three types of traffic: constant bit rate, variable bit rate, and available bit rate.

- *Constant bit rate (CBR).* CBR traffic includes transmissions such as voice and video traffic. To handle this type of traffic, the ATM network can be configured to act like a dedicated circuit. It provides a sustained amount of bandwidth, low latency, and low cell delay variation.

- *Variable bit rate (VBR).* VBR traffic is handled similarly to CBR, except that the bandwidth requirement is not constant. For example, an ATM network supporting a video conferencing application guarantees bandwidth during the video conference. During the video conference, however, the actual amount of bandwidth used can vary.

- *Available bit rate (ABR).* ABR traffic does not require a specific amount of bandwidth or specific delay parameters and is quite acceptable for many of today's applications. Applications such as e-mail and file transfers are usually supported by ABR connections. LAN emulation and TCP will also use ABR connections. Figure 3.16 shows why and illustrates how link capacity is allocated to each different traffic type. While CBR reserves a constant amount of the total available bandwidth, VBR requires that a large amount of spare capacity be available. ABR defines a way to use this very valuable spare capacity. It can provide service that is no worse, and is in many cases better, than most of today's networks, but it requires only a very low amount of bandwidth.

ATM networks are usually architected to provide the performance guarantees required by CBR and VBR traffic and to allow ABR traffic to

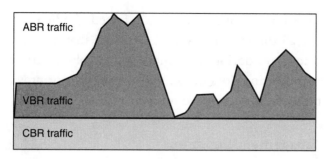

Figure 3.16 *Link usage by different traffic types.*

use the remaining bandwidth. Table 3.3 summarizes the three types of traffic and their network requirements.

Establishing ATM Network Connections

To establish an ATM connection, the ATM end station (i.e., the calling party) asks the ATM network for a connection to another ATM end station (i.e., the called party or destination) by initiating a connection request. The connection request leads to a negotiation between the calling party and the ATM network (a connection establishment procedure). The parameters being negotiated are specified by the ATM Forum User-Network Interface (UNI) Version 3.0, and include sustained and peak bandwidth, burst length, and QoS class. As a result of this connection establishment procedure, a contract is secured between the ATM network and the ATM end station. The ATM network promises to deliver a guaranteed QoS, and the ATM end station promises not to send more traffic than it requested in the connection establishment procedure.

Table 3.3 Summary of Network Traffic Types and Their Requirements

Traffic Type	Example	Bandwidth Required	Cell Delay Variation	Latency
Constant	Voice, circuit emulation	Guaranteed	Minimal	Low
Variable	Compressed video	Guaranteed	Variable	Low
Available	Data	Not guaranteed	Wide variation	Moderate to high

To have meaning, contracts must be enforced. Traffic management functions include all the techniques used to ensure that users receive the QoS guaranteed to them during the connection establishment procedure. Further, when congestion does occur, traffic management provides the mechanism that allows the network to recover.

Traffic Management Functions

ATM networks use three techniques to manage traffic: traffic shaping, traffic policing, and congestion control.

Traffic Shaping   Traffic shaping is a management function performed at the user-network interface of the ATM network. It ensures that traffic matches the contract negotiated between the user and the network during connection establishment. Traffic is shaped according to the Generic Cell Rate Algorithm (GCRA), as specified by the ATM Forum standard UNI version 3.0. Devices implementing traffic shaping are typically those connected to an ATM network and include ATM network adapters in PCs or workstations, hubs, bridges, routers, and DSUs.

Traffic Policing   Traffic policing is a management function performed by the ATM network (i.e., ATM switches) and ensures that traffic on each connection remains within the parameters negotiated at connection establishment. To police traffic, ATM switches use a buffering technique called *leaky bucket*. It is a system in which traffic flows (leaks) out of a buffer (bucket) at a constant rate (the negotiated rate), regardless of how fast it flows into the buffer. The need for policing occurs when traffic flow exceeds the negotiated rate and the buffer overflows. The ATM switches must then take action to control (police) it. In the header of each ATM cell is a bit called the *CLP bit*. The ATM switches use this bit to identify cells as either conforming (to the contract) or nonconforming. If cells are nonconforming—for example, if there are more cells than the contract allows—the ATM switch sets the CLP bit to 1. This cell may now be transferred through the network, but only if the current network capacity is sufficient. If not, the cell is discarded and must be retransmitted by the sending device. CBR traffic requires a single buffer (leaky bucket) to police the traffic because CBR traffic uses only a sustained (average) rate parameter in its network contract. VBR traffic uses two buffers (dual leaky buckets) to monitor both the sustained rate over a discrete time period and the maximum (peak) bandwidth used during the connection. If either value exceeds the contract parameters, the ATM switch polices the VBR traffic by manipulating the CLP bit.

Congestion Control In a well-designed ATM LAN, CBR and VBR, traffic experiences the low latency service negotiated at set up, while ABR traffic might experience congestion, depending upon the current loading of the network. Since the applications that use ABR connections are less sensitive to delay, all applications run as planned. Congestion control is needed because the ABR traffic is likely to experience congestion at some point in time. If the congestion is controlled, the ABR service still provides value. CBR and VBR are designed to require no complex congestion management schemes. However, the complete congestion control specification is still being defined by the ATM Forum. Several schemes have been proposed and the two leading schemes involve controlling traffic flow on either a link-by-link basis or an end-to-end basis. ATM Forum is working on a solution that will integrate the best features of each.

- *End-to-end.* End-to-end flow control is readily available from most vendors at a relatively low price. But it has two major drawbacks. First, recent studies show that cells can be lost during congestion. Second, it requires a lot of buffer space. End-to-end schemes control the transmission rate at the edge of the network—where the LAN meets the ATM device. If the ATM-to-LAN connection is Ethernet to the WAN, at which point an access device will convert your traffic to ATM, then this is a low-cost method of connecting to the WAN. Because little of the LAN beyond the backbone uses ATM technology at this time, precise control may not be required, and the few extra buffers will not substantially increase the cost. As ATM equipment prices continue to drop, ATM technology will be used in more of the LAN. Paying for the extra buffer space that will be needed will be expensive, making precise control over more complex networks a very important issue—thus, the compromise to include the option to provide more precise link-by-link control.

- *Link-by-link.* Link-by-link flow control supports more users but uses less buffer space, all without losing cells. Link-by-link has two major drawbacks: It will be more expensive at first, and equipment that implements link-by-link is not yet available. Link-by-link schemes, beside providing control on a per-link basis, also control each VC separately. This allows the network of ATM switches to control a particularly active device, while other devices continue to receive a fair share of the available network capacity. This control is more precise and can be implemented in either of two ways: a rate-based method or a credit-based method. The *rate-based method* controls the flow of traffic by communicating to the sending device the rate at which it can transmit (the allowed rate). The *credit-based method* controls traffic flow by communicating to the sending device the remaining buffer space (credits) that the downstream device (receiver) has available to receive traffic, on a per-VC and per-link basis.

- *Integrated.* The integrated proposal considered by the ATM Forum provides an end-to-end, rate-based scheme as the default method, with the link-by-link scheme as an option. Because most equipment can provide this level of control immediately, users will have a standards-based ABR congestion-control scheme very quickly. If the network requires a more precise congestion-management scheme, the optional scheme can be used to increase the control of the ABR resources. This would still be a standards-based solution, with end-to-end and link-by-link equipment coexisting in the same network. When a connection is made from an end-to-end device, the link-by-link device would simply perform the end-to-end control scheme when talking to that device. This would preserve the existing investment in equipment, while providing for future growth.

Switching Technology

This section discusses the architecture, techniques, and requirements of switching technology.

Broadband Switching Architectures

There are several basic choices a switch designer can make in designing a broadband switching fabric, including architecture, control, and physical topology. A tree-based taxonomy of broadband switching systems design has been used [1]. The goal in classifying switch designs in this way was to group together switches with fundamentally similar designs and to separate those based on radically different philosophies of architecture. In their tree structures, the end nodes represent particular switch designs and each inner node specifies a design decision. This collection of design points classifies broadband switches through their fundamental properties. Five design points were chosen. In descending order on the classification tree, they are as follows:

1. *Dedicated versus shared links.* A link is *dedicated* if at most one switch element may transmit onto it; otherwise the link is *shared*.

2. *Transport mode.* The transport mode is the discipline governing the transmission of data units onto links. It may be statistical or assignment-based. (This corresponds roughly to circuit switching versus packet switching.)

3. *Centralized versus decentralized routing control.* Routing decisions can be made at one point in the switch, or distributed throughout the switch.

4. *Buffered versus nonbuffered switch elements.* A buffered switch element stores data units within the element to resolve output link contention. Unbuffered switch elements incur a constant delay on data units relative to congestion in the system.

5. *Switching technique.* The switching technique is the method of directing the data units from one switch element to another. The four techniques are space switching, time switching, frequency switching, and address filtration.

Dedicated versus Shared Links

The majority of ATM products on the market or in development today use dedicated links, although a few products employ a shared link (e.g., bus) switching approach.

Transport Mode

A transport mode is the method the switch uses to transmit data over links in the interconnection fabric. This mode can be categorized into links whose capacities may be dedicated to each call, or may be made statistically available to all calls through a need-based scheme. Dedicated links form the basis of the connection switching that has dominated the telecommunications business to the present day.

Distribution of Routing Intelligence

With centralized routing, a single logical entity controls the routing of data over the switch elements and interconnection fabric at a given level. In distributed (decentralized) routing, data routing is executed via processing that is localized to each switch element, possibly with the aid of global state information replicated among each of the elements. A *self-routing* switch fabric would be considered to have distributed control. An example of distributed routing is a Batcher/Banyan switch fabric, where the routing control is decided on an element-by-element basis as the data unit traverses the switch. Distributed routing is, in general, more complex than centralized routing. This complexity includes routing control for multiple services, deciding between the arbitration of access of cells to a link and the switching of data. Output link contention occurs when data from one or more switch elements contend for the transmission capacity of a link.

Buffering in Switch Elements

A buffered switch element stores cells within the element to resolve output link contention. Data may be buffered before switching (*input buffering*), after switching (*output buffering*), within the switch elements (*shared buffering*) or not at all (*no buffers*). The choice of buffering method can have a dramatic effect on the characteristics of a switch (e.g., performance under heavy traffic loads).

Switching Techniques

The switching technique influences or determines the physical structure, distribution, and execution of the switch intelligence. There are five primary switching techniques, as follows:

- *Space switching.* The transmitting switch element chooses among multiple physical routes to send data. An example of space switching is a simple crossbar switch, in which a crosspoint either routes a data unit along a vertical output bus or ignores it and allows it to continue along the horizontal input bus.

- *Time switching.* The transmitting switch element delays a data unit for an amount of time dependent on the routing.

- *Frequency switching.* The transmitting switch element chooses among two or more carrier frequencies upon which to route data units.

- *Address filtration.* The receiving switch element selects labeled data units from among multidestination traffic. Fujitsu's FETEX 150 ATM switch architecture [7] is an example of such an arrangement. In the FETEX 150, buffered intelligent crosspoints select cells through addressing, and cells are injected onto horizontal input buses that later transmit the cells onto the attached vertical output buses. The switch elements do not route data units among physical links, but instead select cells from the broadcast input bus and retransmit onto a single output bus.

- *Shared memory.* Cell headers are examined and the cells are pointed to output queue buffers based on destination VCs (per-VC queuing) and on traffic type (e.g., CBR). For each traffic type there may be a small number of dedicated output buffers. When those buffers are full, the excess incoming calls are buffered into a common shared memory pool queue (hence, the name). The output buffers are drained using some equitable technique (e.g., fair-weighted round-robin) designed to support QoS requirements across the switch. Fore's ASX-1000 is an example of such an architecture, as discussed later in this chapter.

These five switching techniques may be used alone or, more commonly, in combinations.

Generic Switch Requirements

Following the current thinking, this discussion encompasses multiservice ATM switches. However, some think that ATM should not be bogged down with supporting legacy services, such as SMDS and frame relay. These observers argue that they have seen this happen before, with the same disappointing results: Develop a new "next-generation" technology, supposedly supporting brand-new enterprise applications—and then, immediately, loose the focus on forward momentum and try to cannibalize existing markets served by previous-generation technologies. Often, this simply results in a watering down of the advanced features of the new technology, to its ultimate detriment. Also, this results in missed price-point goals for the new technology, since it is unable to compete within the optimization envelope reached by the previous technology in support of the legacy requirements. In our opinion, new technology should strive to open up yet-unopened vistas, where the old technology could not ever compete with the new. Some examples in support of this argument follow.

Packet switching was developed as a next-generation technology. Then, instead of focusing on supporting e-mail, point-of-sale, EDI, multicasting, and so on, it tried to compete with SNA private lines as a replacement of multidrop links. It never made it.

Satellite communications was developed as an advanced technology. Then, it immediately attempted to compete with point-to-point terrestrial lines. It never made it. Only when the technology moved beyond that, to very small aperture terminals, did it go somewhere.

ISDN was developed as a multimedia transmission system under an advanced signaling mechanism. Then, it attempted to: (1) compete with dedicated lines and cluster controllers, and (2) support a decade-older technology—namely, packet switching. Up to now, ISDN has failed to impact the data market. Even for Internet access, the market is looking beyond ISDN to ADSL/SDSL.

Another effort was made to support a better brand of packet services—namely, frame relay—under the auspices of ISDN. Access to packet services via ISDN did not go anywhere. (One can almost say that ISDN went nowhere.) As for other brands of packet services, one could just observe if one now gets frame relay under ISDN.

Hence, in view of these experiences, ATM should not repeat the same mistakes. ATM should not look back at legacy systems, it should look forward to new video, image, and multimedia services. ATM should not be bogged down to support frame relay and SMDS under its auspices, but it should focus on being so good, so cost-effective, so feature-rich, that users are enticed to jump ship from frame relay and go expressly to ATM. Why, for example, spend $1 billion (as an imaginary number) to make ATM support a transitional, limited service, rather than opting to spend that $1 billion in making ATM proper even better?

As discussed elsewhere in this chapter, major standardization efforts have taken place during the past decade with regard to ATM. As applied to ATM switches, such efforts have principally focused on the *outer layer* of the switch—namely, the interfaces to the rest of the world. Specifically, for all three planes of interest—user plane, control plane, and management plane—UNI interfaces and NNI interfaces have been defined. The ATM UNI interface is specified in ITU-T, ATIS T1, and ATM Forum documents. The ATM B-ICI[6] (which is a specific realization of the NNI) has been similarly defined by these standards bodies. The support of services other than cell relay service (e.g., frame relay, SMDS, and circuit emulation) requires additional interface descriptions to the user as well as to the network side. Such a user-side interface could be the service-specific interface (e.g., SMDS' SNI); on the network side, the interface to be supported could be to a service-specific network (e.g., SMDS), or it could be to an ATM-based network, if the carrier decides to provide the service over an end-to-end ATM infrastructure.

These interfaces are typically defined up to the ATM layer. When and if the switch provides service interworking (e.g., frame relay in, cell relay

---

[6] In principle, there are two types of NNI interfaces: the Broadband Interswitching System Interface (B-ISSI), which applies between switches of the same carrier, and the Broadband Intercarrier Interface (B-ICI), which applies between switches of different carriers. However, to reduce the number of distinct interfaces, emphasis has been placed by the industry on the B-ICI. In addition, a somewhat distinct NNI that is applicable between private switches has evolved.

out), then the interfaces have to be defined up to the AAL layer (AAL functions are normally supported in CPE).

Issues that are more *internal* to the switch include switching fabric (e.g., shared memory), performance, call admission control, traffic and congestion control, and a large set of operations functions (some operations functions also have interface-level visibility, e.g., flow of maintenance cells). Figure 3.17 depicts a view of an ATM switch. The figure shows three layers: (1) the interface layer, (2) the services and functions layer, and (3) the switching fabric layer. The *interface layer* is the most visible, in the sense that it supports the various UNI, NNI, and operations interfaces in peer fashion with external entities. The *services and functions layer* defines the functional capabilities of the switch, including transport services, signaling processing, routing, and operations. Most switches support similar interface and services and functions layers. The *fabric layer* is where most switches diverge: This layer embodies the cell buffering, switching, queuing, prioritization, and discard disciplines. Unlike time-space-time switching in digital voice, where the switching fabric layer is nearly identical across all vendors, in ATM vendors have choosen fairly distinct paths to achieve the switching function.

The interface layer is typically implemented in hardware, specifically in swappable line cards. There are cards for various UNIs, NNIs, and operations systems. The services and functions may be partially implemented in the inner portion of the line card hardware (e.g., a clock recovery function for a circuit emulation interworking function); however, most of these functions are implemented in firmware and in software. The switching

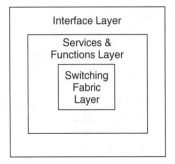

Figure 3.17 ATM switching layers.

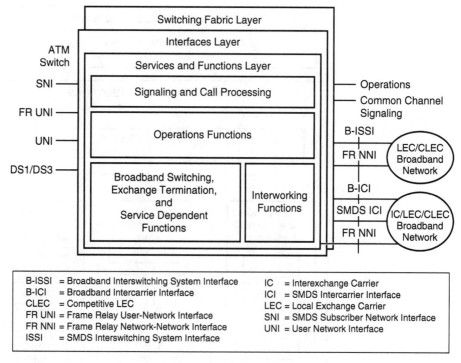

Figure 3.18 *Detailed elements of an ATM switch.*

fabric layer is implemented almost completely in hardware; there may be some form of fabric control that is implemented in firmware.

Figure 3.18 depicts a more detailed view of an ATM switch; in Figure 3.19 some of the interfaces and functions have been identified. Table 3.4 summarizes key areas where requirements have been published to guide vendors in their development.

At the hardware level, an ATM switch may be implemented in a centralized or a distributed form. For example, a distributed switch could include remotely located line termination and multiplexing hardware. These hardware elements would then be connected via umbilical links to the main module in the central office. This is a traditional view of how a local carrier can place functionality closer to the user. With the advent of ATM switches, however, the situation appears to have changed to a new paradigm. Since switches are fairly inexpensive, local carriers tend to locate

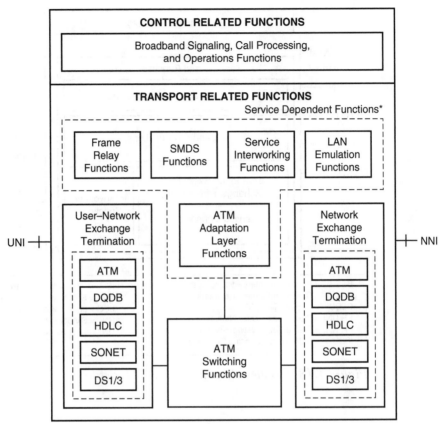

Figure 3.19 *Switch functions.*

lower-end ATM switches[7] (called *edge switches*) on customer premises,[8] particularly in commercial buildings, and connect these edge switches to the
more powerful *core switch* located at the central office via the B-ICI or
B-ISSI interface.

Another kind of distributed switching computing—namely, the positioning of some portions of the switch control at some remote hub location
(central office) while leaving the low-end switching fabric function at the
local central offices—has not been met with interest on the part of the ATM
switch manufacturers, although it has been advocated by some parties.

---

[7] Although such a switch would have fewer functions than a full-fledged switch, it is
still an ATM switch on its own merit.

[8] On the network side of the interface.

Table 3.4 Highlights of ATM Switch Requirements

Support of appropriate physical interfaces
  Support of a multiservice UNI
  Support of a multiservice B-ICI

Support of broadband services
  Support of PVC and SVC cell relay service,* along with appropriate, fully
    functional UNI and B-ICI interfaces in the user, control, and management
    planes
  Support of other fast-packet services (SMDS† and SVC/PVC frame relay‡)
  Support of circuit emulation§ (DS1 and DS3)
  Support of video, voice, and multimedia services

Support of appropriate QoS, at least for reference loads, and related support of
  traffic management

Performance

Reliability

Support of OAM&P, including billing

Spatial and environmental criteria

---

*Cell relay service is a cell-based transport service that enables users to have direct access to the ATM layer capabilities, at a variety of transfer rates. Connections are either Virtual Channel Connections or Virtual Path Connections; the latter are logical groups of Virtual Channel Connections that require identical routing. The service supports constant bit rate, variable bit rate, and available bit rate connections. Initially, the PVC (provisioned) version of the service was available; SVC service became available at a later date.

†SMDS is a public connectionless packet-switched service that provides for the transfer of variable-length data units at high speed. A *connectionless service* is one where the data units are transmitted without first going through call establishment procedures. The data units can be transferred from a single source to a single destination (singlecast), or from a single source to multiple destinations (multicast). Features include source address validation, address screening, and group addressing. A high degree of reliability is guaranteed with respect to the integrity of the delivered data units. SMDS is specified such that subscriber protocols and applications ride on top of the SMDS layer. Initially the carrier's switch may support SMDS via service-specific access interface and service-specific and/or multiservice (integrated) trunk interfaces; eventually, SMDS will be supported on an integrated UNI and B-ICI.

‡Frame relay (ANSI T1.606) is a connection-oriented data transport service that provides for the transfer of variable-length data units. It is typically used for LAN interconnection and enterprise-network applications. Frame relay requires establishment of end-to-end connections (PVC or SVC) before information can be transferred. SVC requires call setup procedures. Customer access to the frame relay service on at ATM platform/switch will initially be over service-specific UNIs.

§This service supports the transport of constant bit rate signals using an ATM infrastructure. Typical examples are nx64 kbps, DS1, and DS3. These bit-oriented periodic signals are generated by user's equipment which employs AAL 1; ATM cells are then generated and transported across the network. AAL 1 provides segmentation and reassembly, as well as clock recovery functions. If there is a need to interwork at the remote end with a non-ATM user, the switch must provide the interworking by appropriately implementing the AAL capability.

ATM Switch Example: Fore's
ForeRunner ASX-1000

This section provides an example of an advanced ATM switch architecture. It discusses the Fore ASX-1000's internal architecture.[9] This section explains the advantages of shared memory switching and buffering, and briefly discusses some of its advantages. Many ATM switches today utilize shared memory switching methods; queue management is, however, what distinguishes one vendor from another.

Shared memory switching architectures are becoming more prevalent in LAN ATM switches. The flexible nature of RAM allows buffers to be manipulated in countless ways—each having certain advantages. This section also details the functionality, scalability, and other aspects of Fore's distributed shared memory switching architecture, using the first-generation monolithic shared memory switching architecture as a point of comparison.

ATM can provide a single network for simultaneous use by both real-time and non-real-time traffic. From the beginning, ATM designers have envisioned a technology that would integrate the world's networks and enable many new networking services. While total network convergence is still in the future, Fore has already begun the migration. This migration allows many of today's ATM users to start moving away from separate networks for data, voice, and video (each of which must be cost-justified and individually managed) to one universal network.

To fulfill this promise, an ATM network must not only provide large amounts of bandwidth, but also must add strict control over QoS, including latency and jitter, at the application level. With enough control, an ATM switch can simultaneously handle all four ATM Forum service classes: CBR, VBR, UBR, and ABR.

With these demanding requirements in mind, only a *nonblocking* switch can offer the necessary performance. (*Nonblocking* means that the switching mechanism can handle the aggregate capacity of all input ports without any possibility of cell loss. Buffering may have to occur on the output

---

[9] This section is based on Fore's White Paper, *ForeRunner ATM Switch Architecture* (April 1996), and is used with permission.

side due to output port congestion, but intelligent buffering systems minimize any negative effects of buffering.) Prospective ATM users must also look beyond performance and consider reliability, scalability, and upgradability (i.e., continuous investment protection). *Reliability* has obvious importance in any production network. *Scalability* addresses future network needs, and *upgradability* keeps pace with the rapid advancements in ATM technology.

Fore's *distributed shared memory* switching architecture provides a fully modular, hot-upgradable, low-latency switching capability that scales easily from 2.5 to 10 Gbps today, and will scale beyond 80 Gbps in the near future.

Fore's Distributed Shared Memory ATM Switching Architecture

Fore's distributed shared memory architecture integrates its highly-refined time division multiplexing (TDM) fabric with its hot-swappable shared memory modules. Using TDM for distribution to the shared memories offers many appealing features. The basic architecture requires no input or internal buffering, makes high-performance multicast implementations simple, and, unlike a monolithic or first-generation shared memory switch, offers redundancy, scalability, and upgradability.

In the *ForeRunner* architecture, cells are processed in three steps. At the input, cells enter an ingress port, are sent through two stages of VPI/VCI translation, and are checked for compliance with their traffic contract. The cells are then distributed, using TDM, to multiple shared memory switching outputs.

In TDM, a multiplexing agent allots specific periods of time to each of a set of data inputs to transmit onto a high-speed common bus. The *ForeRunner* cell distribution fabric provides 2.56 Gbps of bandwidth—enough to service all inputs at full line rate with any combination of *ForeRunner* interfaces. This mechanism, combined with the high-speed shared memory modules, assures fully nonblocking performance.

TDM minimizes fabric traffic and eliminates fabric blocking. Only one cell crosses the TDM distribution path at a time, even for multicast connections. Fore's TDM distribution technique, combined with its advanced shared memory technology, minimizes latency and jitter, enables the switch to transfer real-time traffic along with non-real-time traffic, and, unlike any other LAN switch on the market, provides full bandwidth, nonblocking multicasts.

The ASX-200BX supports four distributed shared memory switching and buffering modules, called *network modules* (see Figure 3.20). Series C network modules support between one and six ports.[10] Each switching module is responsible for switching and queuing cells destined for its local ports. This responsibility includes secondary VC translation, advanced buffer management, and multicast duties.

These shared memory switching and buffering modules are hot-swappable—offering unprecedented reliability, ease of use, mean time to repair (MTTR), bandwidth scalability (e.g., upgrade a DS1 link to an OC-12c link—without rebooting the switch), and upgradability. The network modules queue each VC individually (per-VC queuing), allowing the switch to apply unique and advanced buffer management techniques like weighted-round-robin queuing and early packet discard/partial packet discard (EPD/PPD).

Scalable Distributed Shared Memory

ATM promises scalability—which is considered very important, because computer networking bandwidth demands are increasing so rapidly. The latest in graphics, publishing, telecommunications, and even medical applications software (or any of a thousand other networked applications) is quickly eating up any "extra" bandwidth.

Key to getting the most out of a network investment is the ability to add nonblocking bandwidth to a network quickly, inexpensively, and easily. Another major plus is the ability to add bandwidth without disrupting other network services. The inherent modularity of a distributed implementation leads easily to hot-scalable shared memory switches.

---

[10] The *ForeRunner* Series C Network Modules are Fore's third generation of interface cards. They have powerful custom ASICs managing 13K cells of shared memory with various innovative buffer management techniques.

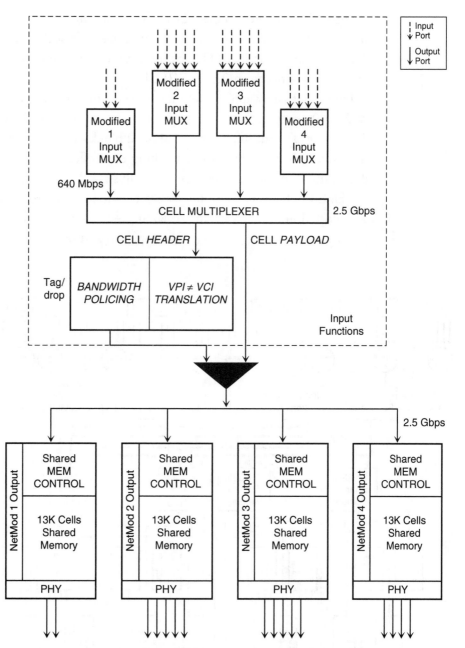

Figure 3.20 ForeRunner *ASX-200WG/BX distributed shared memory switching architecture.*

With this in mind, Fore has leveraged its distributed shared memory architecture to create a switch architecture that scales beyond 80 Gbps. The *ForeRunner* ASX-1000 backbone switch implements this architecture today, scaling between 2.5 and 10 Gbps.

Figure 3.21 may appear daunting. It is a representation of the *ForeRunner* ASX-1000 switch. But it can be broken down into just three different sections: a set of input functions, a 16K-cell block of cell-switching RAM and an associated shared memory switching controller, and up to 4 distributed shared memory outputs per fabric. As illustrated in Figure 3.21, the fabrics are highly self-similar, leading easily to modular, incremental expansion. The dotted lines delineate the scalability—the switch scales from 2.5 to 10 Gbps and the fabrics can be hot-swapped in and out of the ASX-1000.

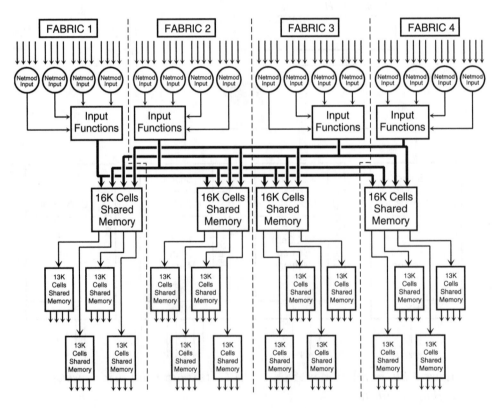

Figure 3.21 ForeRunner *ASX-1000—a 10-Gbps distributed shared memory switch.*

Cells pass through the input functions in the same three-step order as in the ASX–200BX. The major output-side difference is the presence of a two-tier hierarchical switching and buffering system. Cells enter an ingress port, are sent through two stages of VPI/VCI translation, and are checked for compliance with their traffic contract.

Traffic is distributed from the inputs to the appropriate 16K-cell first-tier cell switching and buffering memory via as many as 4 individual 2.56 Gbps unidirectional TDM distribution fabrics. Each fabric transmits on one bus and receives traffic (destined for its output ports) from the input sections of other fabrics on the other three buses. These distribution buses use the same TDM technology as the ASX–200WG/BX. There are simply more of them—as many as four in a fully populated *ForeRunner* ASX–1000.

Once the cell reaches the appropriate 16K-cell shared memory(s), the cell is queued in one of 5 queue sets, 1 queue set per network module, with 1 queue set for cells destined for that fabric's switch control processor. Each queue set consists of four individual priorities to isolate high-priority traffic from low-priority traffic.

The cells drain from the 16K-cell switching fabric memory to the second-tier shared memory switching and buffering modules (i.e., the network modules) and, from there, are transmitted to the network.

Nonblocking Architecture

How is this architecture nonblocking? Remember, *nonblocking* means that the switching mechanism can handle the aggregate capacity of all input ports without any possibility of cell loss. There is the possibility of having to buffer on the output side, but congestion on one port will not affect any other port.

The maximum possible port bandwidth capacity for a 4-fabric ASX–1000 is either 16 OC–12c or 64 OC–3c ports, which are equal in bandwidth—10 Gbps.

As shown in Figure 3.22, each input section of the fabric has a 2.5-Gbps time division multiplexer with which to send cells to any of the 4 first-tier switching/buffering memories. Each 16K-cell first-tier switching and buffering memory can receive as much as 10 Gbps of traffic. This is a max-

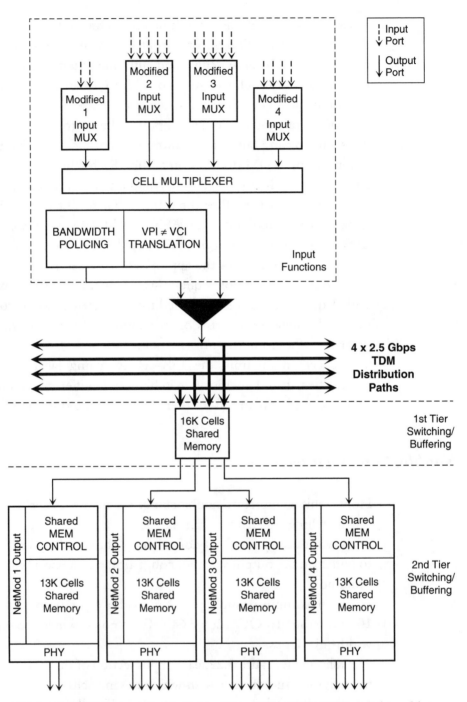

Figure 3.22 *Diagram of a single* ForeRunner *ASX-1000 switching fabric.*

imum of 2.5 Gbps from the inputs on each of the 4 fabrics. The first-tier switching and buffering memory drains at an aggregate 2.5 Gbps to as many as 4 second-tier switching and buffering memories simultaneously.

Therefore, the output side of any one fabric is capable of receiving and queuing the aggregate of all the inputs on all four fabrics, and is thus non-blocking.

Fore's Design

Fore's design utilizes shared memory switching, per-VC queuing, and multicast capability.

Shared Memory Switching

Developments in ASIC technology have allowed Fore to apply highly sophisticated hardware-based, queue controlling ASICs to ATM. Rather than using FIFOs, these ASICs control a bank of RAM, sharing the memory resources among a number of ports. Implemented properly, these shared memory switching ASICs offer many features.

Statistical Advantages   Because all ports have access to more total buffering than their own "fair share," the effective usage of buffer space increases over that of a buffer system comprised of multiple smaller, non-shared buffers.

Bandwidth Management   Shared memory queues do not have the first-in, first-out limitation of FIFOs. Advanced shared memory designs offer crucial capabilities like per-VC queuing, packet level discard, weighted-round-robin queuing, and weighted fair queuing.

ForeRunner ATM switches take advantage of shared memory queue control systems to perform the switching functions. In Fore's case, a cell enters the switch, passes through switch support functions (e.g., VPI/VCI translation and bandwidth policing), and is immediately distributed to a shared memory switching module.

Queue control hardware switches and buffers the cell in that module's bank of RAM. The flexibility inherent in advanced RAM technology allows the sophisticated *ForeRunner* switches to offer unprecedented functionality and expandability.

Monolithic versus Distributed Shared Memory Switching

The first-generation shared memory architecture provides a single controller chipset and a single block of shared memory. One of the key advantages of a shared memory architecture is its ability to share memory across a number of switch ports. This leverage assumes that all ports will not incur congestion simultaneously, providing statistically more bandwidth for each port to use (see Figure 3.23, Table 3.5, and Table 3.6).

The more ports each switch can share, the better its statistical advantage. The question is, at what point do ports stop sharing the bandwidth? Is there a point of diminishing returns?

There are three key factors in this decision. First, the statistical advantage of shared memory tapers off rapidly as more ports are added. For example, increasing the number of ports sharing the space from 1 to 4 gives a 35.3 percent improvement in buffer utilization. Increasing the number from 4 to 8 provides an additional 12.4 percent improvement, but increasing from 8 to 16 grants only a 7.1 percent improvement (see Figure 3.24; References [8] and [9].

Second, reliability plays a large part in a decision to place a certain switch in a network over another switch. A monolithic shared memory implementation typically has a number of *single points of failure,* where the failure of a single component can cause the entire switch to fail. These include the core switching RAM and core switching ASICs. The switch control processor and controller RAM usually also present reliability concerns in a monolithic switch, with no support for redundant control processors or memory.

Third, and most important, modularity provides reliability, scalability, and upgradability. Fore has determined—with much input from its customers—that the best tradeoff lies in the 4- to 8-port range.

Virtual FIFO versus per-VC Queuing

In virtual FIFO queuing, each port has a dedicated set of virtual queues. Incoming cells are placed by the switch at the end of the appropriate queue. Cells pass through the queue in a first-in, first-out fashion. Although this

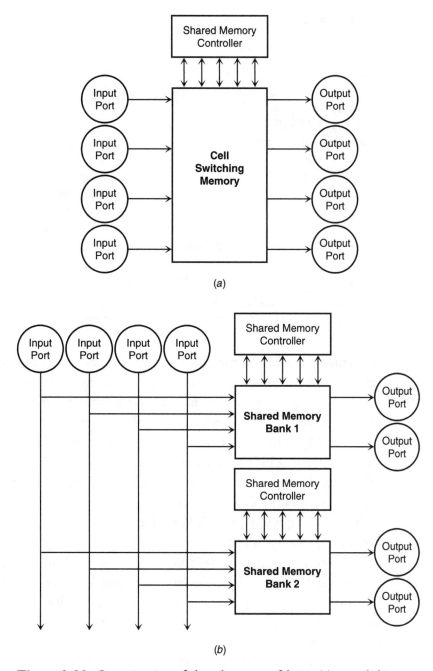

Figure 3.23 *Organization of shared memory fabrics: (a) monolithic
shared memory, and (b) distributed shared memory.*

Table 3.5 Monolithic Shared Memory

ADVANTAGES	DISADVANTAGES
Large, inexpensive buffers.	Single switching memory is a single point of failure.
Can support packet level discard (EPD/PPD).	Core switching functionality is not hot-upgradable.★
High buffer effectivity multiplier.	No upgrade path for additional switch bandwidth.
Can support per-VC queuing.	Typically, implemented with only one processor, with no redundancy or upgrade path for either the processor or the processor memory.★
Can support weighted-round-robin queuing.	See note.★
Can support a per-VC rate controller.	See note.★

★Just because the technology allows for implementation of advanced buffer management algorithms, don't assume that any given shared memory switch will support them—few do.

implementation lowers the cost per byte of buffer from that of true FIFO memory, offers greater aggregate buffer space than FIFO-based switches, and provides a more efficient use of the available buffer space, it does not take advantage of the flexibility of shared memory switching.

Per-VC queuing maintains the features of virtual FIFOs without the FIFO limits. In a FIFO, once the switch adds a cell to the queue, the cell's relative position is fixed. FIFOs, virtual or not, limits buffering flexibility and can introduce VC-level blocking.

In per-VC queuing, each VC is given its own queue. Each VC queue consists of a string of cells, where each cell in the string points to the next cell. These individual queues can then be drained in any order desired, enhancing fairness and minimizing potential VC-level contention. Furthermore, these VC queues can also be read by multiple ports (i.e., all of the ports attached to that shared memory), dramatically reducing buffer load during multicast.

Per-VC queuing enables weighted-round-robin queuing methods, and these queue-draining techniques lower jitter and latency far below those of FIFO-based techniques.

Table 3.6 Fore's Distributed Shared Memory

FEATURE	BENEFIT
Distributed memory eliminates a single point of failure.	Increased network uptime.
High buffer effectivity multiplier.	More efficient buffer utilization.
Early packet discard/packet level discard (EPD/PPD).	High (80%) "goodput" during congestion so bad as to cause 1% cell loss.
	Prevents TCP from breaking down under cell loss conditions.
Per-VC queuing.	Crucial for weighted-round-robin queuing.
	Crucial for explicit rate (ER) ABR flow control.
	Crucial for nonblocking, efficient multicast in a shared memory switch.
Largest aggregate buffering space of any ATM LAN switch (272K cells).	Absorbs 10-Gbps bursts of network traffic without losing any cells.
Upgradable buffering.	Eases network expansion.
Weighted-round-robin queuing.	Improves network performance.
Hot-swappable switching fabrics.	Eliminates a single point of failure.
Bandwidth of ASX-1000 is hot-upgradable in 2.5-Gbps increments from 2.5 to 10 Gbps.	Provides upgrade path for *nonblocking* bandwidth—without disrupting network services.
Modular processors.	Simplifies future upgrades of switch control processing power and memory.
Network modules.	Provides hot-upgradable core switching functionality and switch buffering to protect technology investment.

The Series C network modules' shared memories implement per-VC queuing. These queues are drained using weighted round-robin. Per-VC queuing enables many other advanced features, such as early packet discard and partial packet discard, and provides more control over QoS parameters.

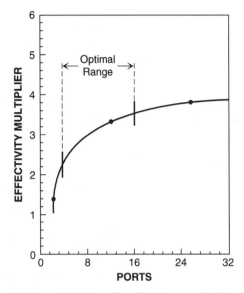

Figure 3.24 *Buffer effectivity multiplier.*

Multicast Capability

Without per-VC queuing, shared memory multicast operations require a cell copying mechanism. Two limitations exist in these implementations: the number of cell copies that the mechanism can make per cell time and the amount of buffer space necessary to hold multiple copies of cells in the same memory space.

Fore's Series C Network Modules—with per-VC queuing—use a patented nonblocking multicast method. FORE switches append a *scorecard* to each cell, identifying which destination ports need to transmit the cell. Each port, after it has serviced a particular cell, removes its identifier from the scorecard and moves to the next cell as directed by the pointer.

If its identifier was the last identifier in the scorecard, the port tells the shared memory controller to delete the cell. This scorecard implementation of multicast maximizes multicast performance, allowing *ForeRunner* ATM switches to provide nonblocking multicast performance.

This innovative implementation also minimizes the number of copies of the cell existing in memory at any given time. In a 16-port monolithic shared memory switch that implements virtual-FIFO queues (with associated multicast copying mechanism), a broadcast would cause 16 copies of the cell to exist in the memory simultaneously. Fore's per-VC queuing leads to a maximum of 1 cell per network module, for a maximum of 4 cell copies in the 16–port ASX–200BX.

Conclusion

Shared memory switches offer flexibility far beyond that of other switching architectures. Distributed shared memory switches, like those of the *ForeRunner* product line, offer the following clear advantages over monolithic shared memory switches:

- Easy upgradability
- Reliability
- Scalability
- Advanced buffer management

Advanced buffer management techniques include per-VC queuing, packet level discard, and weighted-round-robin queuing.

References

1. D. Minoli and M. Vitella. *Cell Relay Service and ATM for Corporate Environments.* New York: McGraw-Hill, 1994.

2. D. Minoli and T. Golway. *Designing and Managing ATM Networks.* Greenwich, Conn.: Manning/Prentice Hall, 1997.

3. D. Minoli and A. Alles. *LAN, ATM, and LAN Emulation Technologies.* Norwood, Mass.: Artech House, 1997.

4. D. Minoli and A. Schmidt. *MPOA.* Greenwich, Conn.: Manning, 1998.

5. Peter Newman. "ATM Local Area Network." *IEEE Communications Magazine* (March 1994): 86 ff.

6. D. Minoli and G. Dobrowski. *Principles of Signaling For Cell Relay and Frame Relay.* Norwood, Mass.: Artech House, 1995.

7. S. Nojima, E. Tsutsui, H. Kukuda, and M. Hashimoto, "Integrated Services Packet Network Using Bus Matrix Switch," *IEEE Journal on Selected Areas in Communications* 5 (8): 1284–1292.

8. Y. Shobatake, et al. "A One-Chip Scaleable 8 ★ 8 ATM Switch LSI Employing Shared Buffer Architecture." *IEEE Journal on Selected Areas in Communications* 9 (Oct. 1991): 1248–1254.

9. H. Kuwahara, et al. "A Shared Buffer Memory Switch For An ATM Exchange." IEEE CH2655-9/89/0000-0118: 4.4.1–4.4.5

4 LAN Emulation and Classical IP over ATM

This chapter discusses methods that can be utilized to support interworking of legacy and ATM LAN networks, utilizing switched technologies. The ATM Forum's LANE specification allows deployment of ATM networks in such a manner that connectivity is retained across technologies. The goal of LANE is to simplify attaching legacy devices to an ATM network in a manner that is no more difficult than making attachments to an Ethernet or Token Ring. The key to such connectivity is to utilize existing network layer protocols (e.g., IP and IPX) on the ATM network.[1]

Three related internetworking technologies that make use of switched services are covered in this chapter:

- The ATM Forum's LAN Emulation
- Use of ATM as a fat pipe
- The Internet Engineering Task Force (IETF)'s Classical IP Over ATM (CIOA)

Each approach has strengths and weaknesses. These factors need to be taken into consideration by network managers when implementing a migration strategy from existing LAN infrastructures to ATM-based LAN internetworks. Most internetworking vendors today support all three of these approaches. As discussed in the following chapters, vendors are also looking to support MPOA, as well as new approaches, such as IP switching.

---

[1] Portions of this chapter are extended from References [1] and [2].

Rationale for LAN Emulation

In an effort to facilitate the deployment of ATM technology, the ATM Forum has developed specifications to connect the installed base of LANs to newly deployed ATM-attached end systems.[2] The interconnection needs to be done such that the LAN-attached stations and servers have no knowledge that they are communicating with ATM-attached devices. Also, applications running on ATM end systems need to be unaware that they are using ATM as a network technology.

With LANE, the ATM Forum has developed a technique for emulating the behavior of legacy LAN protocols and carrying the data over an ATM network. LANE provides a mechanism that enables ATM-connected end systems to establish MAC–layer connections to legacy LANs. The resulting hybrid infrastructure is called an *emulated LAN* (ELAN); Figure 4.1 shows an example. LANE utilizes the ATM backbone to achieve wire-speed con-

---

[2] These are also known as *end stations.*

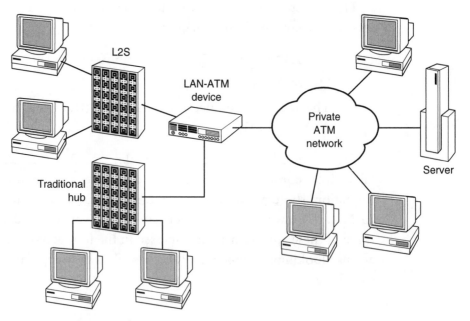

Figure 4.1 *ELAN layout.*

nectivity among devices. LANE permits several ELANs to concurrently share the same ATM network. Typically, such an ATM network is a private campus network. Hence, LANE's goal is to solve the problem of interconnecting an organization's installed base of LAN devices over a new ATM media, while minimizing the impact to existing systems. For example, in a client/server environment LANE provides a mechanism for existing LAN-based client/server applications to run over ATM networks without modification. Table 4.1 summarizes key features of LANE.

Ethernet and Token Ring are designed to operate in a *connectionless* manner. On connectionless networks, any end system can communicate with any other end system simply by placing a packet onto the network (e.g., see Figure 4.2). The packet is transmitted to all end systems attached to the network by virtue of the fact that they all share a common virtual "wire." Sending data to all end systems simultaneously is called *broadcasting*. Figure 4.3 highlights some of the key features of LANs.

By contrast, ATM is a *connection-oriented* service. With a connection-oriented network, a path between end systems must be established by an appropriate apparatus before the end systems can communicate (see Figure 4.4). Transmitting a message to all end systems on the network via broadcast is not possible with ATM because it would require that an explicit connection be made with all end systems on the network—this is usually impractical. Furthermore, because virtual circuits provide for one-to-one communication, an end system that is broadcasting to a set of other end systems must be aware of any additions or deletions to the set. Figures 4.5 and 4.6 depict how broadcasting and multicasting can be achieved in LANs and in ATM services. As illustrated in Figure 4.6, a set of point-to-multipoint

Table 4.1 LAN Emulation

Utilizes a classical bridging model that supports MAC-layer connectivity
Reduces broadcast traffic
Supports multiple ELANs
Supports dynamic configuration
Improves network security
Provides bandwidth management
Supports existing LANs
Supports Spanning Tree

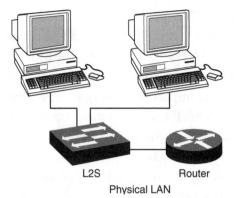

L2S   Router
Physical LAN

Figure 4.2 *Typical LAN layout.*

connections must be established, maintained, and administered in order to achieve multicasting. The ATM switch must support cell replication. To achieve broadcast, the multicast group must be equal to the entire population (this means setting up a point-to-multipoint connection to all relevant network devices); to achieve any-to-any communication, a complete set of full-community point-to-multipoint connections must be established.

As discussed in Chapter 3, *permanent virtual circuits* (PVCs) are created by a management interface to the ATM switch and exist until they are deleted

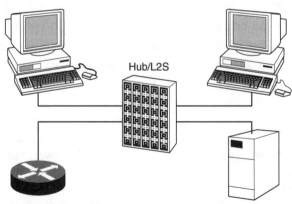

- Connectionless technology.
- Every data unit (packet) has a destination and source address associated with it.
- Variable packet size—max of 1500 bytes.
- Each host has a 6-byte MAC address.

Figure 4.3 *Traditional Ethernet.*

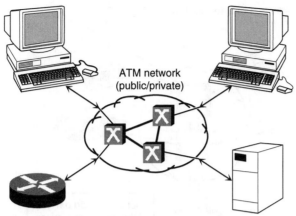

ATM network
(public/private)

• Connection-oriented technology.
• Establishes a Virtual Channel Connection (VCC) first, then
  sends data units over the VCC.
• Fixed data unit size—53 byte cells.
• Each ATM host has a 20-byte NSAP address.

Figure 4.4 *ATM networks.*

by the network administrator. Virtual circuits that are dynamically created
and deleted are called *switched virtual circuits* (SVCs). However, SVCs require
additional capabilities in each end system and/or switch. In addition, SVCs
introduce problems such as call routing, call control, and billing to end systems so that call setup messages can be correctly formatted.

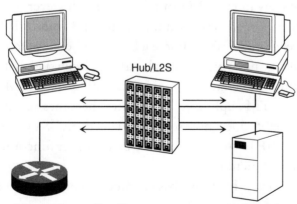

Hub/L2S

• Achieved by specific addresses.
• Broadcast address FF:FF:FF:FF:FF:FF.
• Multicast groups can be set up with a Multicast group address.

Figure 4.5 *Ethernet broadcast/multicast.*

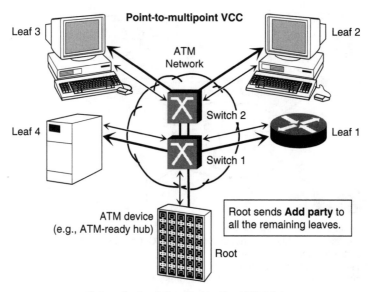

Figure 4.6 *ATM broadcast/multicast.*

ATM networks of the early to mid–1990s only supported PVCs. Higher-layer protocols, like HTTP, Telnet, and FTP, were intentionally kept unaware of the fact that they were using ATM as a connectivity service. In many ATM networks, routers have been attached to the network by HDLC-to-ATM interworking converters. These frame-to-ATM conversions effectively hide ATM from the router. However, as the ATM network becomes larger, establishing more and more PVCs becomes administratively undesirable. For each pair of end systems needing to exchange data, a new bidirectional virtual circuit needs to be created. While this can be tolerated for networks with a few ATM devices (e.g., routers with ATM uplinks), it quickly becomes a difficult design approach, since it does not scale up effectively. SVC techniques are required by LANE to support both the data transfer functions and some of the initialization functions.

LANE provides a bridging-based internetworking solution with the following characteristics:

- Based on the ATM Forum UNI 3.0/3.1 specification
- Provides a high-performance, scalable backbone

- Can provide protocol-independent switching across logical LANs
- Can support PVCs, SVCs, or combinations
- Supports autoconfiguration of the network

The goal of LANE is to make possible the attachment of legacy devices to the ATM network in a manner that is no more difficult than making attachments to an Ethernet or Token Ring network. LANE allows the addition of end systems or bridges to the ATM network.

The LANE protocols can also be extended to operate across wide area connections. By extending LANE over the WAN, users have the capability of transparently attaching remote end systems to the various campus ATM LANE networks. With LANE, end systems that are part of an ELAN over a wide area connection are treated as end systems directly connected to the campus network.

Based on its design, all devices attached to a LANE system can function in a plug-and-play fashion, with minimal manual configuration. Because user application programs using the network services will interface with MAC device drivers, the LANE service does offer the same MAC driver service interface to upper protocol layers. To accomplish this, LANE client software must be loaded onto end systems that are attached to an ATM. The LANE protocol stack (see Figure 4.7) provides all the functions of com-

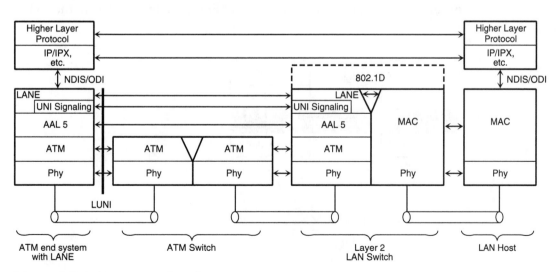

Figure 4.7 *LANE protocol stack.*

monly used device driver APIs, NDIS, and ODI. The problem of hiding ATM from applications is solved by replacing the NDIS device driver with LANE. The new device driver hides ATM by recreating the traditional API that is used between applications and Ethernet or Token Ring device drivers. Note in Figure 4.7 that the UNI-based interface between ATM devices running the LANE software is called *LANE UNI,* or *LUNI.* Figure 4.8 provides a graphical view of what LANE accomplishes, while Figure 4.9 depicts a more physical view of the implementation.

The key elements of the LANE architecture were introduced in Chapter 1, and they are summarized in Tables 4.2 through 4.7.

Example of LANE Operation

Table 4.8 depicts the key LANE processes. Initialization is a fairly complex process; however, this should take place only when a device is turned up or moves to a new location on the virtual LAN.

When an ATM end system running LEC software (e.g., a UNIX end system, PC, Macintosh, switch processor, or LAN access device) becomes active on the network, it automatically locates the LECS either by the Interim Local Management Interface (ILMI), by the well-known address, or on a Configuration Direct PVC. For private networks, the ATM

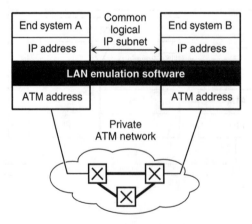

Figure 4.8 *LAN Emulation.*

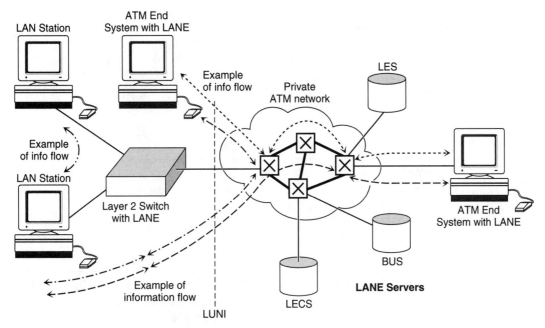

Figure 4.9 *LANE protocol interfaces.*

Table 4.2 LANE Components

LANE servers.
LAN Emulation Clients (LECs).
LUNI interface specifies the operations between the LEC and the network
  providing the LANE services.

Table 4.3 LANE Servers

SERVER	FUNCTION
LAN Emulation Configuration Server (LECS)	Provides configuration information.
LAN Emulation Server (LES)	Implements control functions for a particular ELAN.
Broadcast and Unknown Server (BUS)	Performs all broadcasting and multicasting functions.

Table 4.4 LANE Configuration Server (LECS)

Maintains configuration information in the ATM network.
LECS is the first point of contact when new clients boot and need to register on their ATM network.
LECS is the entity that assigns LANE clients to particular ELANs.

Table 4.5 LAN Emulation Server (LES)

Implements the control function for an ELAN.
Provides mapping of ATM identifiers to a MAC identifier.
Only one logical LES per ELAN.

Table 4.6 LANE Broadcast and Unknown Server (BUS)

BUS is used to flood unknown destination address and forward multicast and broadcast traffic in an ATM connection-oriented network.
Each LEC is associated with a single bus per ELAN.

address is based on the NSAP format, as shown in Figure 4.10. Figure 4.11 depicts the ILMI operation.

Figure 4.12 shows the logical interactions required to support LANE, while Table 4.9 summarizes the LANE connections required for initialization and information transfer. To provide a sense of the required (logical) connectivity, Figure 4.13 shows the various VCCs overlaid on the LANE servers and LECs.

Table 4.7 LAN Emulation Client (LEC)

Performs data forwarding, address resolution, and other control functions to emulate Ethernet or Token Ring.

Table 4.8 LANE Processes

Initialization on the ATM network.
Registration with the LANE server (LES).
Registration with the BUS server.
Data transfer with other LANE clients.

- LANE uses SVC.
- SVC needs addressing.
- Two types of addressing: E.164 and NSAP-like.
- LANE uses private network addressing (i.e., NSAP-like).

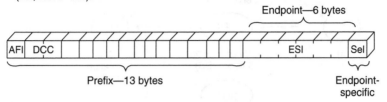

Figure 4.10 *ATM address.*

The ELAN concept can be utilized to support virtual LANs (VLANs). In the sections that follow, VLAN is utilized as a realization of the ELAN concept.

A Client Joins a VLAN/ELAN

The LECS provides a list of VLANs to which that end system may belong.[3] In the simplest case, if the end system is a member of one VLAN,

---

[3] This section is based in part on promotional material from Fore Systems, Pittsburgh, Pa. Used with permission.

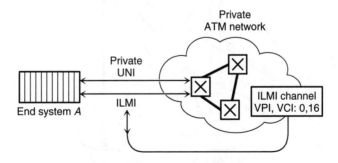

- End system *A* has a preassigned End Station Identifier (ESI—6 bytes).
- The 13-byte ATM address prefix is obtained from the ATM switch via ILMI.
- ILMI uses the SNMP request/response mechanism over ATM.

Figure 4.11 *Interim Layer Management Interface (ILMI) autoconfiguration.*

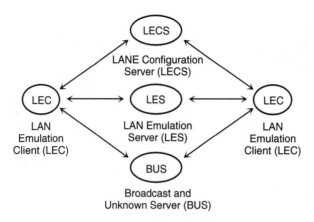

Figure 4.12 LAN Emulation logical interactions.

there is one inquiry and one response. The end system disconnects from the LECS and joins the VLAN.

In a more complex case, the end system is a member of multiple VLANs and there are several iterations of inquiries and responses to discover the VLANs to which the end system belongs. The LECS also returns the address of the LES for each of the VLANs of which that ATM end system

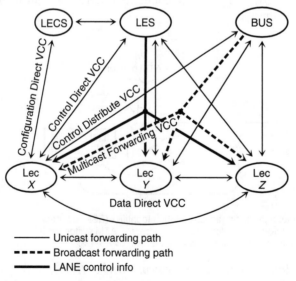

Figure 4.13 LANE initialization paths.

Table 4.9 Virtual Channel Connections (VCCs) to Support LANE

VCC	FUNCTION
Configure Direct VCC (LECS)	Used by the LEC to obtain the LES address for the ELAN it wishes to join.
Control Direct VCC (LES)	Used by the LEC to join the ELAN and for LE-ARP requests.
Control Distribute VCC (LES)	Used by the LES to forward unresolved LE-ARP requests.
Multicast Send VCC (BUS)	Used by the LEC to send broadcast and unresolved unicast traffic.
Multicast Forward VCC (BUS)	Used by the BUS to forward broadcast and unresolved unicast traffic received on the Multicast Forward VCC to the remaining LECs.
Data Direct VCC (LEC-LEC)	Used for the bulk of data traffic.

is a member. The ATM end system starts a LEC instance for each VLAN and joins those VLANs by contacting the LES for each VLAN.

The LEC sends an ARP to the LES with a MAC address of hexadecimal FF. The LES provides the LEC with the address for the BUS. The LEC now has all the necessary LAN Emulation Services information to actively participate in the VLAN.

Data Transfers over a VLAN/ELAN

As with 802.x LANs, each end system in the ATM LAN has a unique MAC-layer address. When ATM end system *A* needs to send data to ATM end system *B*, *A* first looks for the MAC address for end system *B*.

If the MAC address–to–ATM address translation is not already in its cache, *A* determines the ATM address of *B* by sending an address resolution message to the LES. Since any ATM-connected end system must register with the LES, the LES can directly answer *A*, providing the ATM address of *B*, which is contained in the LES cache. This is illustrated in Figure 4.14, Steps 1 and 2.

Once *A* discovers the ATM address for *B*, any existing ATM LANE connection between them can be used (e.g., permanent or switched virtual cir-

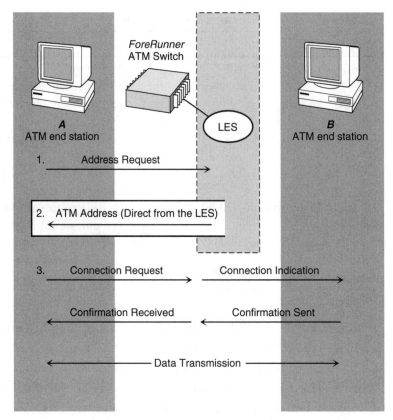

Figure 4.14 *ATM end station* A *discovers ATM end station* B *and transfers data to* B.

cuit). If no connection exists, *A* initiates a connection using ATM signaling (i.e., Q.2931 signaling protocol). This is shown in Figure 4.14, Step 3.

Multicast and Unknown Traffic

Multicast traffic is any message that uses a multicast MAC address (i.e., a broadcast, group, or functional MAC address, where the destination address of the 802.3 frame is all hexadecimal FF). Multicast traffic and unknown traffic are handled by the BUS.

The most common multicast address is the *broadcast address.* Any message sent to the broadcast address, by definition, must be sent to every station in the VLAN. When the BUS receives a broadcast request, it transmits that message to every member of the VLAN. Multicast traffic capabilities support true multicast applications, such as distance learning, multipoint video conferencing, and protocols (e.g., NetBIOS, IP, and IPX).

Unknown traffic is any data for which the sender has not yet obtained an ATM address. For example, end systems on different LAN segments (e.g., via a transparent bridge) will not receive the ATM address resolution messages. The transparent bridge answers such inquiries. Once the bridge learns of the remote end system, it will answer the address resolution messages and a direct connection will be used.

BUS Example: ATM-to-Ethernet

Figures 4.15*a* and 4.15*b* describe the sequence of events that occurs when end system *A* on the ATM LAN wants to discover end system *X* on the Ethernet LAN and then transfer data to *X*. First, *A* sends an address request message to the LES (see Figure 4.15*a*, Step 1). The request is forwarded to the LAN access device and two outcomes are possible:

- The LAN access device, acting as a proxy LEC for *X*, knows the address and provides its own ATM address in response to the request (see Figure 4.15*a*, Step 2*a*).
- The LAN access device does not know the address of X and sends no response.

As *A* sends an address request message to the LES, it simultaneously sends the data to the BUS (See Figure 4.15*b*, Step 2*b*). The BUS sends (i.e., floods) the data to all end systems on the VLAN. When the data reaches the LAN access device (previously registered as a proxy LEC with the LES), it is broadcast to all end systems on the Ethernet LAN. *X* receives the data and then responds to the LAN access device. The LAN access device stores the MAC address of end system X in its table.

When *A* receives no response to its initial request, it resends the address request message. (See Figure 4.15*b*, Step 2*c*.) This time, when the LAN access device receives the request, it has the MAC address of *X* in its table. So, the LAN access device can reply, sending its ATM address to *A*. *A* can now send its data to the LAN access device, which forwards the data to *X*.

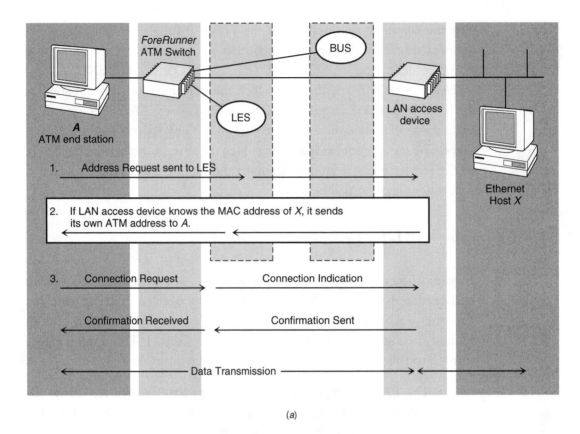

(a)

Figure 4.15 (a) ATM end station A discovers Ethernet end station X and transfers data to X. The LAN access device knows the MAC address of X. (b) The LAN access device must discover the MAC address of X.

Once A knows the ATM address of the proxy LEC, a connection to the LAN access device is established, if one does not already exist. Data transmission begins with the LAN access device sending data directly to end system X (see Figure 4.15a, Step 3, or Figure 4.15b, Step 3).

How Existing LAN Protocols Operate over ATM

Most existing LANs are connectionless systems using bridges or routers to expand the size of the network. They reduce internetwork traffic by dividing the network into smaller, more manageable pieces called *segments*. These bridges and routers use broadcast techniques that send every message on a network segment to every end system on that segment.

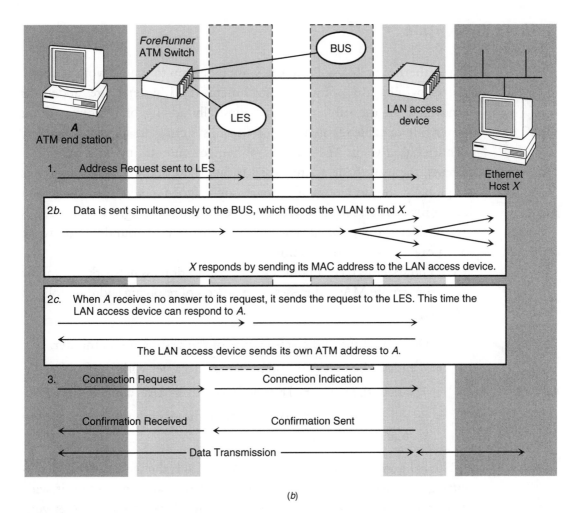

(b)

Figure 4.15 *(Continued).*

In contrast, ATM networking uses a connection-oriented scheme, as previously discussed. Data sent from one device to another on an ATM network is seen only by the destination device. ATM networks use two types of connections—PVCs and SVCs. *PVCs* are manually configured by the network manager. *SVCs* are created dynamically by ATM switches, using signaling software.

ATM networks and existing LANs differ in another way—address structures. ATM networks use a 20-byte OSI NSAP address; existing LANs, a 48-bit MAC address.

Address Resolution

In an Ethernet or Token Ring LAN, membership to the network is determined by physical points of attachment. End systems are added or removed from the subnetwork by physically adding or removing them from a hub or cable segment. Once an end system is attached to a LAN, it identifies itself via its *MAC address*. When communicating with other end systems, the most basic address used is the MAC address. MAC addresses are globally unique; blocks of addresses are assigned to vendors to be coded into computers and routers at the time of manufacturing. MAC addresses are considered to be *flat addresses*, because they can be found in a network in a more or less random fashion.

For end systems to communicate on a connectionless network, they must locate the remote MAC addresses. The process of resolving addresses follows these steps:

- Determine the end system name of the destination computer.
- Translate the text end system name into a network layer address.
- Translate the network layer address into a MAC address.
- Establish communication to the remote end system using the acquired MAC address.

The Address Resolution Protocol (ARP) [3] is the technique for transmitting a request onto the network to ascertain if any end system is willing to accept packets containing the destination Layer 3 address. Network layer addresses, or Layer 3 addresses such as IP, are typically used to identify end systems on a network in a somewhat human-readable format. Layer 3 addresses are also used to build a hierarchical view of the network. The hierarchical view facilitates routing and locating end systems.

ARP packets are transmitted to the broadcast address on the network. The broadcast address allows the ARP request to be forwarded to all end systems on the subnetwork. On an Ethernet, the broadcast address is all 1s. Any packet containing all 1s as a destination address will be accepted by all end systems on the subnetwork and its contents will be examined. An ARP packet contains the source MAC address of the client attempting to locate the server and it also contains the requested destination address.

ARP broadcast packets are forwarded through bridges and hubs, but typically will not pass through routers [3].

When an end system receives an ARP packet, it examines the contents to determine if the source of the ARP is attempting to locate itself or some other end system. If the destination Layer 3 address matches the end system's (i.e., the *self check* is true), it responds to the ARP, then notifies the ARP originator that it has successfully located the destination. If the requested address does not match the end system's Layer 3 address, then the ARP requests are ignored [3].

When a client has received a successful reply to an ARP message, it can update its internal ARP table with a Layer 3–to–MAC address mapping. The ARP table is referenced when the client has additional packets to transmit. If an entry for the server exists in the client's local ARP cache, then the packet is immediately transmitted. If no ARP entry exists, the address resolution process will be repeated. ARP table entries have a limited lifetime and, if not used, are periodically deleted to allow for changing Layer 3 addresses.

ARP was designed for connectionless networks. When end systems are placed on a connection-oriented ATM network, they lose the ability to transparently broadcast to all end systems. Special techniques need to be used to mimic connectionless behavior. To overcome the lack of broadcast capability, LANE provides the ARP function via the LAN Emulation Configuration Server (which provides configuration information), the LAN Emulation Server (which implements address registration and resolution) and the Broadcast and Unknown Server (which performs broadcasting and multicasting functions).

ATM LANE handles the connections transparently and implements an address resolution procedure to accommodate the different addressing schemes. LANE provides a mechanism for end systems to obtain a mapping between conventional 48-bit MAC addresses and 20-byte NSAP ATM addresses. Data transmitted across the LUNI is converted from packets to cells and vice-versa, using the AAL5 to perform the SAR function—segmentation (i.e., packets to cells) and reassembly (i.e., cells to packets; see Figure 4.16). Figure 4.17 shows this flow of data from an ATM end system to an Ethernet end system. (In this example, both end systems are using IP or IPX as the network layer protocol.) LANE soft-

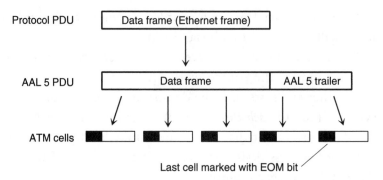

Figure 4.16 *ATM Adaptation Layer 5 (AAL5).*

ware on the ATM end system receives an Ethernet PDU and passes it to the ATM adapter in the end system.

The LAN access device recognizes that the data must be sent to an Ethernet client and reassembles the cells into packets in the AAL SAR on the LAN access switch. Next, LANE software transforms the data into an Ethernet PDU and sends it to the Ethernet end system.

Transparent Operation

So that existing LANs and ATM networks can interoperate, Ethernet or Token Ring frames are transferred using LANE software. This allows two things: carrier-type LAN access devices, linking LANs to ATM networks, and support for existing operating systems and protocols designed for 802.x LANs.

While routers routinely span different LAN media, the ability to use bridges to do the same is important in many LANE environments. By connecting existing LANs with bridges, new stations connected to the ATM internetwork appear as if they are still part of the same preexisting LAN. This means that applications that previously worked with all stations on the same LAN will continue to work correctly, even though some stations are now directly connected to the ATM internetwork— LANE allows the ATM segment to be treated as just another LAN segment.

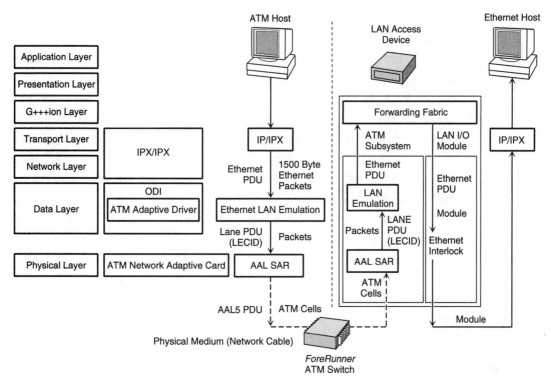

Figure 4.17 *Data flow from an ATM host to an Ethernet host via a LAN access device.*

Application Example: E-mail

E-mail is a familiar application that clearly illustrates the benefits of LANE. Figure 4.1 shows a typical corporate network consisting of both ATM end systems and Ethernet end systems. The e-mail server is attached to an ATM switch in the network, as are various end systems. Servers are placed on the ATM network to take advantage of ATM's performance and quality of service. Ethernet end systems are also connected to the ATM network via a LAN access device.

Users who communicate most frequently with one another are grouped within their own VLAN. LANE allows the creation of multiple VLANs/ELANs on the ATM internetwork. These users would typically be grouped based on department or functional group (e.g., accounting, engineering, or marketing).

Although users in the accounting department communicate with one another on a daily basis, they also need to communicate with others in the company. If the e-mail server on the ATM network is a member of several or all VLANs, it could distribute e-mail within one VLAN, as well as between VLANs.

Further, the ability to group users into separate VLANs is an issue for network administrators using legacy LAN technology. With LANE and the LECS, network administrators can easily define VLANs—regardless of physical location. End system computers can be dynamically added to or removed from a VLAN.

Using LANE, end systems on the Ethernet segment can also be grouped into a VLAN, or with end systems on the ATM network. The seamless migration between the Ethernet LAN and the ATM internetwork provides access to the e-mail server and to other end systems on the network, regardless of the LAN technology being used by those end systems.

Implementation of LANE Servers

Figures 4.18 and 4.19 depict two examples of LANE implementations. The various LANE components—LECS, LES, BUS, and LECs—can be implemented in various places in the network, whether local or wide area.

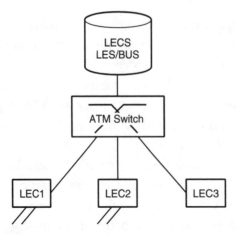

Figure 4.18 *Example of LANE implementation.*

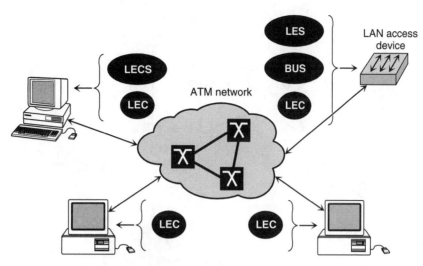

Figure 4.19 *Example of LANE implementation.*

Robust implementations of LANE support multiple, distinct VLANs. Each VLAN acts like a separate network with respect to any other VLAN, and each has it own LES and BUS. Broadcast and unknown traffic is seen only by the end systems connected to a particular VLAN. One end system can to belong to more than one VLAN. To communicate across ELANs, however, an external router is required, as shown in Figure 4.20. There are concerns that this router can become both a bottleneck and a single point of failure.

Limitations of LANE

Emulation of Ethernet and Token Ring is a reasonable migration path if users are interested in migrating their production network to ATM. However, there are consequences, both good and bad, of hiding the ATM network from clients and servers with LANE. By hiding ATM, end systems are not required to understand all of the complexities of operating on a connection-oriented network that supports multiple qualities of service. Conversely, hiding ATM may be suboptimal, since there is no way for an end system to make use of quality of service. (There are no capabilities in LANE for communicating quality-of-service requests, which is one of the fundamental strengths of ATM.) A video server using TCP/IP over ATM

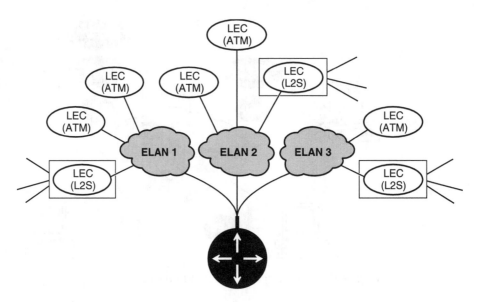

Figure 4.20 *ELAN interconnection.*

is an example of the potential problems encountered due to the lack of quality of service. The question of direct interaction between client/server applications with the ATM network's quality-of-service features has been deferred to LANE Version 2.

While the approach of hiding ATM is pragmatic for quick migration to an ATM network, standalone LANE is a protocol that may have a limited lifetime. It is, however, used within the context of MPOA. LANE, in its most basic form, is just a high-speed replacement for Ethernet. One could argue that 100BaseT would be just as effective for high-speed communication; however, Ethernet-based systems do not support quality of service. Therefore, it is important that protocols that provide access and control of quality of service, like MPOA, be quickly deployed.

LANE also suffers from the limitation that it behaves like protocol-independent bridging. Bridging, while effective for interconnecting small workgroups, does not scale well to large internets. As bridged networks grow they are prone to experience a large load of broadcasts, as end systems try to resolve MAC addresses into Layer 3 address mappings.

It should also be noted that LANE only supports emulation of one type of network at a time. If an end system on an emulated Ethernet wants to communicate with an end system on an emulated Token Ring, the packets must pass through a router that is a member of both emulated LANs.

The LANE Version 1.0 specification does not solve such networking problems as conversion from Token Ring to Ethernet. LANE is limited to the degree that it emulates the legacy network and does not support all functions of the Layer 2 protocol being emulated (the limited support is often due to the paradigm shift from a connectionless network to a connection-oriented one).

On the positive side, LANE supports any Layer 3 protocol that is designed to operate on a connectionless network (e.g., APPN, IPX, IP, AppleTalk, etc.) without modification. Another strength is LANE's support for multicast. Because the protocol emulates connectionless behavior, applications that make use of broadcast or multicast can still function in a LANE network. Also, LANE does not employ collisions or transmission of tokens and beacon frames to transfer information, making it more efficient in that respect.

ATM as a Fat Pipe

There is a simple way to utilize ATM in an enterprise network when ATM is directed at supporting only a backbone function. (This has not been the case in the discussion so far, although one can trivialize LANE to do just that much.) In this application, the router's backbone is replaced with an ATM "cloud." This simply implies upgrading the router to support ATM on the network side. The WAN service is transparent to the rest of the network; the service is just a better, fatter channel, akin to a private line service.

ATM services can be PVC- or SVC-based, as previously described. Services can be CBR, VBR, ABR, or UBR, also as previously described. To use the ATM internetworking service as described in this section, the network manager should follow these steps:

1. Upgrade the router with an ATM uplink.
2. Contact the service provider and set up PVCs in terms of VPIs/VCIs and bandwidth, as well as service class.

3. If SVC service is contemplated, the router must be appropriately configured and the access to the public ATM network must be put in place. (After the access is in place, the router can signal its connection requirements to the network.)

Classical IP over ATM

Classical IP over ATM (CIOA) predates LANE and has been developed by the IETF. It allows a more sophisticated use of ATM than as just a fat pipe insofar as it makes provisions for address resolution within its construct (The fat pipe makes no assumptions about ARP; it must be supported or provided externally.) The IETF specification is defined to provide native IP support over ATM and is documented in RFC 1577 [4] and RFC 1483 [5]. These protocols are designed to treat ATM as "wire," with the special property of being connection-oriented and therefore requiring a means for address resolution. Table 4.10 provides a snapshot of what CIOA is about.

In CIOA, the ATM switching infrastructure interconnecting a group of end systems is considered to be one *logical IP subnetwork* (LIS). The concept behind CIOA is that network administrators will build networks with the same point of view that they use today, partitioning end systems into groups called *subnets,* according to administrative and workgroup domains, and then interconnecting subnets via routers. A LIS in CIOA is comprised of a collection of ATM-attached end systems and ATM-attached IP

Table 4.10 Classical IP over ATM Highlights

Developed by the IETF (RFC 1577; January 1994).
Supports only IP.
No multicast or broadcast support.
RFC 1483 LLC/SNAP encapsulation supports MTU size up to 9180 bytes.
Supports the logical IP subnet (LIS).
    A group of IP nodes that connect to a single ATM network and belong to the same IP subnet.
    ARP Server is used to resolve the ATM addresses of nodes within its LIS.

routers which are part of a common IP subnet. Policy administration, such as security, access controls, routing, and filtering, will still remain a function of routers. Figure 4.21 depicts this arrangement.

Each LIS has an ARP server that maintains IP address–to–ATM address mappings. All members of the LIS register with the ARP server, and, subsequently, all ARP requests from members of the LIS are handled by the ARP server. This mechanism allows direct IP-to-ATM address mappings (see Figure 4.22). IP ARP requests are forwarded from end systems to the LIS ARP server. The ARP server, which can be simply a process running on a ATM attached router, replies with an ATM address. When the ARP request originator receives the ATM address, it can then issue a call setup directly to the destination end system. As opposed to LANE, the CIOA model's simplicity reduces the amount of broadcast traffic and interactions with various servers. By reducing communication with the LECS, LES, and BUS, the time required for address resolution and subsequent data transfer can be greatly reduced. However, the reduction in complexity does come with a reduction in functionality.

CIOA has the drawback that it can only support the IP protocol because the ARP server is only knowledgeable about IP. Communication between LISs must be made with ATM-attached routers that are members of more than one LIS. One physical ATM network can logically be con-

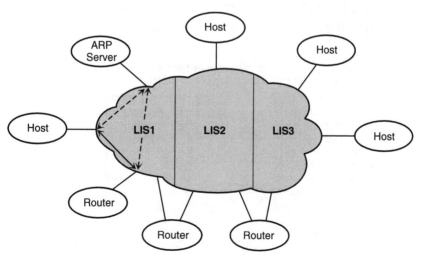

Figure 4.21 *RFC 1577—Classical IP.*

- Each host registers its IP address and ATM address with an ARP server for that logical IP subnet (LIS).

- All hosts ARP directly for ATM addresses using ARP servers.

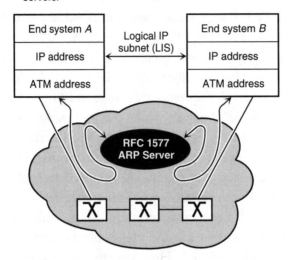

Figure 4.22 *CIOA.*

sidered several LISs, but the interconnection, from the end system's perspective, is via another router. Using an ATM-attached router as the path between subnets prevents ATM-attached end systems in different subnets from creating direct virtual circuits between one another. There are also questions about the reliability of the IP ARP server, in that the specification has no provisions for redundancy.

- After address resolution has occurred, the hosts create a direct VCC for data traffic using UNI signaling.

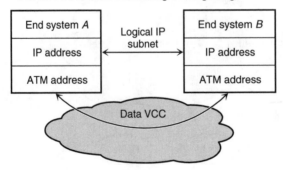

Figure 4.23 *Connections via CIOA.*

In the CIOA model, data transfer is done by creating a VCC between end systems and then using LLC/SNAP encapsulation and AAL5 segmentation and reassembly (see Figure 4.23). Mapping of IP packets to ATM cells is specified in RFC 1483, Multiprotocol Encapsulation over ATM. RFC 1577 specifies two major modifications to traditional connec-

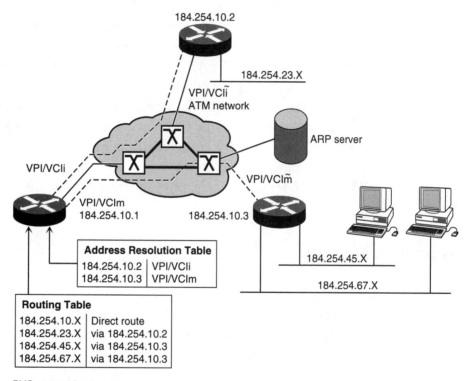

PVC approach

   Step 1. Routing table maps final destination to next hop (e.g., for 184.254.45.g the route is via 184.254.10.3).
   Step 2. Address resolution table or server maps next hop IP address to PVC (e.g., for 184.254.10.3 the VPI/VCI is m).
   Step 3. Forward cellularized packet over ATM virtual connection.

SVC approach

   Step 1. As for PVC.
   Step 2. Address resolution table or server maps next hop IP address into ATM address (E.164 or NSAP).
   Step 3. Signaling stack signals switch over preestablished signaling channel. Network-provided VPI/VCI is placed in table and will be used for transmitting series of cells from original IP PDU.

Figure 4.24 *Example of CIOA operation.*

tionless ARP: The first is the creation of the ATMARP message used to request addresses; the second is the InATMARP message used for inverse address registration.

When a client wishes to initialize itself on a LIS, it establishes a virtual circuit to the server. Once the circuit has been established, the server contains the ATM address of the client and then transmits an InATMARP request in an attempt to determine the IP address of the client that has just created the virtual circuit. The client responds to the InATMARP request with its IP address, and the server uses this information to build its ATMARP table cache. The ATMARP cache is used to answer subsequent ATMARP requests for the client's IP address. Clients wishing to resolve addresses generate ATMARP messages, which are sent to their server, and then locally cache the replies. Client cache table entries expire and must be renewed every 15 minutes; server entries for attached end systems time out after 20 minutes.

Figure 4.24 depicts an example of the operation of CIOA.

References

1. D. Minoli and A. Alles. *LAN, LAN Emulation, and ATM*. Norwood, Mass.: Artech House, 1997

2. D. Minoli and A. Schmidt. *Client/Server Over ATM*. Greenwich, Conn.: Manning/Prentice Hall, 1997.

3. W. R. Stevens. *TCP/IP Illustrated*. Reading, Mass.: Addison-Wesley, 1994.

4. M. Laubach. Request for Comments 1577: Classical IP and ARP over ATM. January 1994.

5. J. Heinanen. Request for Comments 1483: Multiprotocol Encapsulation over ATM Adaptation Layer 5. July 1993.

5 *Multiprotocol over ATM*

The concept of the virtual router is now surfacing for use in enterprise networks, in conjunction with switched communication services. A *virtual router* (which could also be called a *distributed router*) is a collection of devices that in unison present the external behavior of a traditional bridge or router. But being distributed, it has the advantage of optimizing task processing and increasing reliability. Distributed routing is the realization of partitioning the higher-level functions for route determination from the lower-level functions of switching information through a network. A distributed router consists of a central route server that controls multiple edge devices; the edge devices do most of the switching. Routing servers run routing protocols and supply routes to the edge devices. Together the route server and edge devices can be used to build Layer 3 protocol-independent distributed architectures. The evolving MPOA specification describes how these distributed components utilize ATM to provide the routing services in a distributed fashion.[1] MPOA also describes how these services and capabilities are visible to legacy end systems and hosts, and how enterprisewide networks can be constructed with this technology.

As discussed in Chapter 1, the MPOA working group of the ATM Forum has been chartered with developing specifications for application-transparent transmission of legacy Layer 3 protocols, such as IP or Novell's IPX, over ATM backbones.[2] MPOA also allows newer Layer 3 protocols, such as IP's RSVP, to take advantage of ATM's quality-of-service features

---

[1] The authors would like to thank A. Schmidt, Ameritech Data Services, for his insightful input on this chapter.

[2] The specification became available in 1997.

over the same ATM backbone. The goal of MPOA is to ensure that both bridging and routing are preserved for legacy LANs and VLANs already in use in an organization. MPOA utilizes LANE principles, and so provides three key benefits: (1) integration of intelligent VLANs, (2) utilization of cost-effective edge devices, and (3) affording of an evolutionary path for users from LANE to MPOA. MPOA is built from special-purpose servers, from new software on ATM-attached devices, and from the information flows between these components. Table 5.1 describes the benefits of this approach.

While an ELAN's scope in LANE is a single Layer 3 subnet, MPOA is focused on intersubnet connectivity. Furthermore, MPOA aims to enable the separation of the route calculation function from the actual Layer 3 forwarding function in internetworking devices, thereby improving overall performance. In the MPOA model, routers retain their traditional functions, so that they can be the default PDU forwarder and continue to support the forwarding of short flows. Routers also become MPOA/route servers, and provide the Layer 3 forwarding information used by MPOA clients. This allows MPOA clients to set up direct cut-through ATM connections between VLANs, to forward long flows without having to experience extra router hops. The MPOA specification synthesizes the existing LANE bridging work and the IETF Next Hop Routing Protocol (NHRP) work. In a way, the specification provides support for a virtual router with distributed virtual nets or subnets. Table 5.2 provides a summary capsule of MPOA, Table 5.3 enumerates some benefits of MPOA, and Table 5.4 highlights some differences between LANE and MPOA.

Table 5.1 Advantages of MPOA

Users can establish direct connections to remote end stations without using routers.
Provides lower latency in establishing connections between devices.
Provides reduced levels of broadcast traffic.
Allows flexibility in the selection of Maximum Transfer Unit size to optimize performance.
Takes advantage of ATM; direct interdomain connection and a QoS.
Allows for the creation of VLANs on one physical network.

Table 5.2 What MPOA Does

Any Layer 3 protocol can be transmitted across the ATM network without the
   obligatory use of routers and can make QoS requests.
Operates by relying on route servers to maintain knowledge of the location of
   devices—when the location is found, the ATM network can then be used to
   place a call directly between hosts.

This chapter provides a description of MPOA as of press time. Although
the MPOA specification is still under development, many of the key features and concepts are well established. Equipment supporting MPOA has
been available since 1996 [1].

(Note that MPOA work by the ATM Forum took place in 1995 and
1996. The work has had a degree of contentiousness, as vendors jockey for
various positions. In the fall of 1996, after a certain baseline text had been
agreed to and had been around for a relatively long time, most of the text
was scrapped and rewritten.)

Rationale for MPOA

Figure 5.1 depicts a traditional router. Such a router includes interface line
cards, a backplane, and a routing engine. Traditional networks use multiple routers to which subnets are connected. A routing protocol such as
RIP (which is a link-state system) or OSPF (which is a distance vector system) are utilized to propagate connectivity information (see Figure 5.2).

Table 5.3 Benefits of MPOA

Users can establish direct connections to remote servers without using routers.
Provides low latency in establishing connections between devices.
Provides reduced levels of broadcast traffic.
Allows flexibility in selecting MTU size to optimize performance.
Provides the connectivity of a fully routed environment.
Takes advantage of ATM; direct interdomain connections and QoS.
Separates switching from routing.
Provides a unified approach to Layer 3 protocols over ATM.

Table 5.4 MPOA versus LANE

MPOA is an evolution of LANE and uses LANE.
LANE operates at Layer 2: bridging.
MPOA operates at Layer 2 and Layer 3: bridging and routing.
LANE hides ATM/QoS; MPOA exposes both.
LANE requires no modification to host protocol stacks; MPOA requires modi-
fication.

Routers' architectures are now being rethought. MPOA is one of the key technologies that will be considered in the rearchitecting effort.

There are at least two drivers for MPOA in this context, as follows:

1. Build a routed enterprise network that does not utilize the unscalable model of multihop multifacility connectivity between routers, but instead uses a switched communication service, such as ATM. Establishing a direct virtual circuit that crosses multiple administrative domains is a nontrivial problem. This problem is known as the *large cloud problem,* and it stems from the feature set of conventional routing protocols. These routing protocols operate by summarizing or aggregating information to populate their routing tables. As information is summarized, details about the topology are either lost, hidden, or obfuscated. For example, to establish one-hop direct communication across an ATM network,

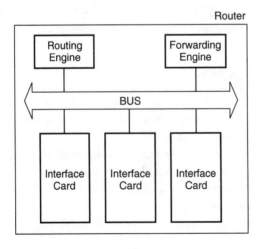

Figure 5.1 Traditional router.

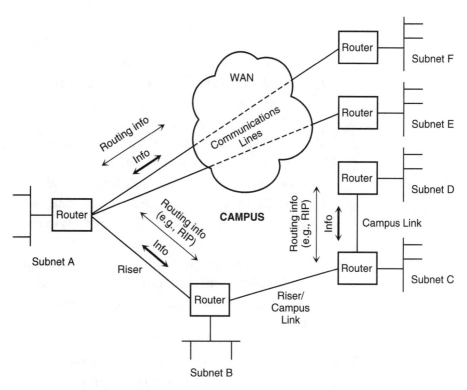

Figure 5.2 *Traditional enterprise network.*

applications need the details of exact locations of destination, not summarizations of various granularity. In addition to the issue of summarized information, multihop communication suffers from problems such as the following:

- Multiple transmission delays through the string of links
- Multiple processing delays through the collection of internet nodes
- Jitter (delay variation) introduced through multiple transmission channels

2. Develop a better VLAN technology. As discussed in Chapter 2, VLANs are being installed both for business flexibility and for improved security. In VLANs the associations and logical addresses of end systems are separated from their physical locations. Com-

mercial VLANs on the market are either portcentric or device-centric. *Portcentric* refers to a set of distributed network ports that are administratively configured to appear as on one common hub. *Device centric* refers to a set of devices that are associated with the computer's network address. In the portcentric model, users are manually assigned to a VLAN, and the ports that make up the VLAN are kept in a database. This model operates as a MAC-layer bridge, forwarding messages between members of the VLAN. Moving a user from one port to another removes this user from its VLAN, requiring the managed VLAN database to be manually updated. In the devicecentric model, end systems are identified by either their MAC or network layer address. Here the network administrator assigns users to a VLAN group, using their addresses as an identifier. Both the location of users and the transport of data between users are managed by the network and are based on the user's address.

Both the enterprisewide network evolution and the VLAN issue benefit from ATM technology. Hence, a mechanism to make effective use of ATM capabilities is sought. For example, using ATM for VLANs is attractive from both a manufacturer's and a network designer's perspective. Manufacturers view VLANs as an opportunity to build a device that is less complex than a conventional multiprotocol router (for example, a LANE bridge is a device that only understands how to register itself on the ATM LAN and communicate with the ATM servers). To a network designer, VLANs are attractive because they provide the ability to invest in one common fabric for the enterprise network. Also, this approach will make VLANs standards-based.

LANE and CIOA (discussed in Chapter 4) can be used to build VLANs. In LANE, a VLAN is equivalent to an ELAN. As noted, inter-ELAN communication requires a router that is a member of two or more ELANs. This router can become a bottleneck. In the CIOA model, a VLAN is a group of end systems aggregated into a LIS; membership in the VLAN is controlled by the ARP server. In the MPOA model, a VLAN is similar to a virtual subnet or virtual network. However, inter-VLAN connections are not necessarily accomplished by routers or routing functions. With

MPOA, end systems are capable of communicating with each other directly, even in the case where the path is between different LISs (without passing data through routers). With MPOA, *any* Layer 3 protocol is able to communicate across the ATM network without using routers, and is able to make quality-of-service requests. The MPOA model operates by relying on route servers to maintain knowledge of the location of devices. When the location is found, the ATM network can then be used to place a call directly between end systems. Figure 5.3 provides a view of what an MPOA-based enterprise network would look like.

One analogy is to view MPOA as a "big router in the cloud" that handles the routing function. It is interesting to note, as a side observation, that computer networking people invariably turn to telephonylike models when they need to scale up to large networks. For example, it was soon learned in the early 1980s that one does not build a LAN with a contiguous user-resident cable, but one uses wiring closets and hubs (a concept that has been used by telephone companies for over 100 years). The general model of MPOA-like routing has been used by the telephone company for decades to support 800 numbers. When an individual dials an 800 number, the serving switch does not know how to route the call to completion. Instead, the call setup is interrupted (for a few tens of ms), until a network-centralized database is queried for the actual (physical) address that corresponds to the logical (800) address, whether that 800 number has no discernible recognition or is some vanity number, such as 1-800-7MINOLI. The local switch is then instructed to cut through the call and complete the connection to the actual destination; all of that takes place in 100 ms or so.

MPOA Components

In the MPOA model, the ATM network is considered to be one physical network capable of supporting many VLANs. An MPOA network consists of several logical components. Logical components are eventually realized in actual hardware, such as router servers, edge connection devices, broadcast servers, and LIS coordinators.

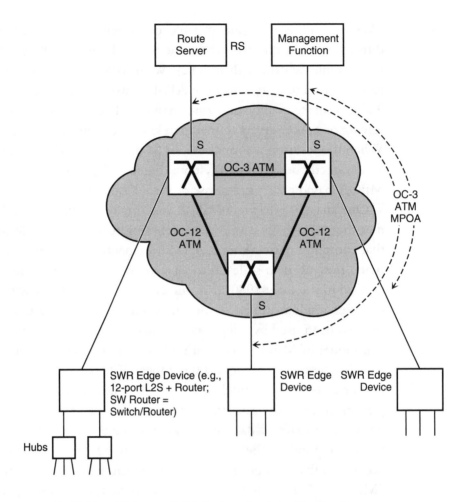

S = Public or Workgroup Switch (replaces bus in traditional router)

SWR = Edge device (e.g., 12-port L2S plus router = switch/router; replaces line cards in traditional router) (switch/router)

RS = Route server (replaces route engine in traditional router)

Figure 5.3 *Implementing enterprise networks with MPOA-based components.*

Functional Groups

Figure 5.4 shows the MPOA system architecture abstraction as described in the baseline MPOA specification [3]. All the MPOA constructs are defined in terms of *functional groups* (FGs). Functional groups represent collections of capabilities that need to be implemented together. The specification describes the required behaviors for each functional group. Abstract functional groups are used, so that the implementer and designer are not constrained as to how or how many of these capabilities are placed in one physical device (the implementation and separation of actual processes running on devices is left to the implementer; MPOA only defines logical components, not products). Functional groups include the following (also see Table 5.5):

- ATM host functional group (AHFG)
- Default forwarder functional group (DFFG)
- Edge device functional group (EDFG)
- IASG coordination functional group (ICFG)
- Internet address subgroup (IASG)
- Remote forwarder functional group (RFFG)
- Route server functional group (RSFG)

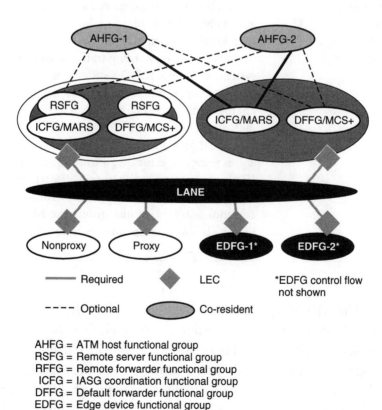

AHFG = ATM host functional group
RSFG = Remote server functional group
RFFG = Remote forwarder functional group
ICFG = IASG coordination functional group
DFFG = Default forwarder functional group
EDFG = Edge device functional group
LANE = LAN emulation
MARS = Multicast Address Resolution Server

Figure 5.4 *MPOA systems architecture.*

Table 5.5 MPOA Functional Groups

GROUP	PURPOSE
Internet address subgroup (IASG)	A functional group that defines the logical space over which MPOA operates.
ATM-attached host functional group (AHFG)	A functional group that is concerned with the set of properties understood by end systems directly connected to the ATM MPOA network.
Default forwarder functional group (DFFG)	A functional group that is responsible for the operation of forwarding data in the absence of direct end system–to–end system connectivity (similar to a LANE BUS).
Edge device functional group (EDFG)	A functional group that is used to connect a legacy LAN to an MPOA network. This device knows how to operate in an MPOA network and screens the legacy LAN devices from MPOA details. These are similar to LANE bridges.
Internetwork address subgroup coordination functional group (ICFG)	A functional group that is needed to support the distribution of a single subnet across multiple legacy LANs and ATM ports (similar to a LES).
Remote forwarder functional group (RFFG)	A functional group tightly coupled with the RSFG and responsible for forwarding traffic from an origination end system, across the MPOA system, and finally to the destination end system, in the absence of a direct virtual circuit. This functional group can be considered to be a global DFFG.
Router server functional group (RSFG)	A functional group that is at the core of MPOA/NHRP and is responsible for distribution of Layer 3 (e.g., IP or IPX) information in the system. It participates within a legacy peering session using legacy routing protocols and communicates the results among the MPOA route servers. It also answers queries to locate end systems (ARP function).

The MPOA specification describes the functions of the FGs and the interactions among them. From a high-level view, MPOA's logical components can be classified as an *MPOA server,* which maintains global knowledge of the Layer 2 and Layer 3 protocols and the topologies for the areas they serve; and an *MPOA client,* which is responsible for maintaining a listing of Layer 2 addresses to Layer 3 for the end system(s) it is communicating with.

The *Internet address subgroup* (IASG) represents a key concept in MPOA. Most internetworking protocols operate in terms of the internet or subnet (e.g., an IP subnet). A similar concept must be supported on ATM; for example, ATM supports the LIS via the RFC 1577 mechanism. An IASG defines the logical space over which MPOA operates. The IASG represents those aspects of the subnet that MPOA supports: It includes the concepts of broadcast scope and of aggregation, which is at the bottom of the routing hierarchy. However, in MPOA, ATM-attached end systems no longer need to be cognizant of the address aggregation. Default forwarding can be used, even between end systems that are in the same IASG.[3] Direct communication can be used between end systems that are in different IASGs. The MPOA system allows connectivity to a single subnet to be distributed across several edge devices. In order to ensure that the collection behaves as a single subnet, LANE-based bridging is used: A logical, virtual-bridged LAN is created to support the subnet, with connectivity on all participating edge devices [1].

The routing protocol and control functions of the virtual router reside in the *route server functional group* (RSFG) and the *remote forwarder functional group* (RFFG). End systems on traditional Ethernet and Token Ring LANs communicate with the RSFG or RFFG as if it was their router.[4] By using bridging for this base communication, a number of goals are achieved: (1) the default path to the RSFG or RFFG exists reliably, (2) communication exists without extensive coordination, and (3) the same mechanisms used to support the distributed subnet also support control and default communication. The RSFG and RFFG provide control functions associated with traditional routers. The RSFG runs the routing protocols and also manages the cache information required by the edge devices; it participates in the NHRP resolution protocol used by all aspects of MPOA for address resolution. The RFFG provides default, connectionless-packet forwarding between IASG. It forwards packets according to the routing information maintained by the RSFG.

---

[3] The support of multiple ELAN segments within an IASG was initially considered by the MPOA group, but it has been deferred.

[4] While the edge device functional group, discussed later, could actively serve as a proxy for the RSFG or RFFG, this would require significant control coordination. In addition, complexity would be introduced when multiple RSFGs or RFFGs were used.

NHRP allows stations to obtain the information needed to establish direct communication (ATM VCs) to communications peers on the ATM network that are not in the same IP subnet. These communications peers may be ATM-attached end systems, or they may be the routers that are the entry and egress to and from ATM for a particular destination. NHRP is used by MPOA for its registration and address resolution needs.

The *ATM host functional group* (AHFG) represents the functions that an ATM-attached end system (which does not wish to use LANE) needs to support to participate in MPOA. By and large, these include registration and the use of the NHRP query and response protocol. The AHFG also may use default forwarding when it does not wish to establish a direct virtual circuit to its communications peer [1].

The *IASG coordination functional group* (ICFG) and the *default forwarder functional group* (DFFG) jointly provide the capabilities required to coordinate AHFG participation in the IASG. The ICFG accepts registrations from MPOA AHFG and answers queries about those end systems. The DFFG provides default forwarding services to those ATM-attached end systems and provides forwarding between the LANE portion of the IASG and the AHFGs.[5]

The MPOA system also encompasses edge devices that are responsible for bridging and cut-through, and for internetworking-level forwarding. These devices logically consist of a bridge, usually supporting multiple VLANs, as shown in Figure 5.5 [1]. Abstractly, the group is known as the *edge device functional group* (EDFG). Hence, the EDFG is a logical layering on top of a bridge. The MPOA edge device is an enhanced bridge. (The bridge is enhanced to understand internetworking PDU formats and to be able to detect flows.) MPOA, in performing a cut-through function, needs a certain amount of details (e.g., ATM addresses) to set up VCs. To solve this problem, MPOA has an associated query protocol. The query protocol is used to follow the routed path to the destination. When the query protocol reaches the final route server, it asks that server for the Layer 2 address of the destination station (or host). When the ATM address is returned, the source can establish a VC that can cut through the

---

[5] ICFG and DFFG elements were initially considered by the MPOA group, but they have been deferred. ATM hosts are still supported in their own subnets [1].

ATM network. NHRP is designed to deal with switched networks such as ATM (but also frame relay, traditional packet switching, etc.). In a network supporting VCs, devices attached to the same network must establish paths to exchange data but, unlike LANs, they lack the ability to easily broadcast a message to all end systems. In the NHRP model, these networks are called *nonbroadcast, multiaccess* (NBMA). The NBMA Next Hop Resolution Protocol allows an end system or router to determine the internetworking layer addresses and NBMA addresses of suitable NBMA next hops toward a destination. If the destination is connected to the NBMA subnetwork, then the NBMA next hop is the destination end system itself. Otherwise, the NBMA next hop is the egress router from the NBMA subnetwork that is nearest to the destination end system. Figure 5.6 shows graphically how cut-through works.

The EDFG has two logical functions: *flow detection* and *flow utilization*. The EDFG must monitor the flow of traffic to MPOA-participating des-

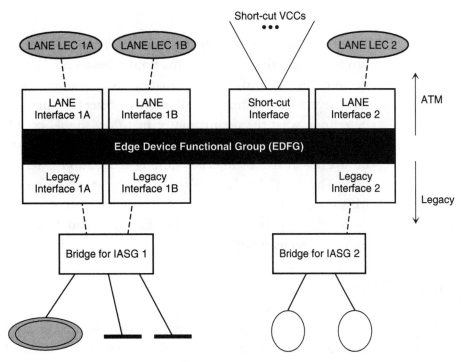

Figure 5.5 Edge device architecture.

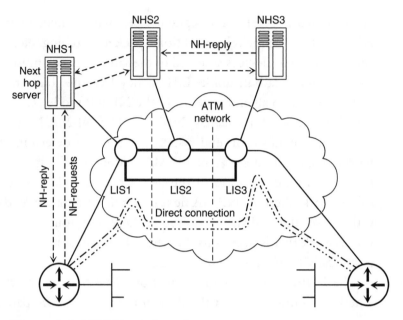

Figure 5.6 *NHRP cut-through.*

tinations. Using LANE, MPOA provides identification, so that an EDFG can recognize which MAC destinations are associated with MPOA. Internetworking traffic to those designations is monitored for flows to specific internetworking destinations. If a flow is detected, then a query is sent to determine what the destination of that flow should be. Once a flow has been detected and resolved, an ATM VC is established. At that time, traffic for that MAC or internetworking destination is redirected over that flow. The MAC headers are stripped as part of the redirection, and the MPOA-specific header is added. The edge device is also an adaptation device that supports Ethernet, fast Ethernet, and Token Ring encapsulation and segmentation into ATM.

MPOA Flows

An MPOA system utilizes a number of information flows between the various FGs. The information flows describe how the FGs exchange MPOA state information. Table 5.6 summarizes these flows.

Table 5.6 MPOA Flows

TYPE	FUNCTION
Client control flows	Used after configuration to query the MPOA servers and inform servers of state change
Configuration flows	Used to retrieve configuration information and register themselves in the network
Data transfer flows	Used to support the actual transmission of information between end systems using an MPOA system
Server control flows	Used between MPOA servers that contain routing information and assist in client-to-client connection establishment

MPOA Operation

As implied by the discussion in the previous subsections, MPOA works by allocating tasks to FGs and then defining the protocols' operation by specifying information flows between the functional groups. Operational steps involved in the MPOA operation cover the following areas:

- Startup and configuration
- Registration and discovery
- End system–to–end system data exchange
- Routing protocol support

These steps are summarized in Table 5.7 [2, 3].

Example of an Edge Device

There are a number of ways to implement an MPOA edge device. Following is a description of an edge device that has been described in the trade press in support of IP networks, based on Reference [1] (this implementation description is only at a high level). This implementation assumes that the bridging is supported by 802.1D. One also needs to

Table 5.7 MPOA Operation

STEP	FUNCTION
Startup and configuration	As in LANE, devices in an MPOA network must contact a configuration server at boot time. The configuration server knows which clients are associated with which virtual networks and notifies the client of its server's ATM address. When route servers and subnet coordination servers (ICFGs) are initialized, they need to be given the identity of the subnets they control, along with the Layer 3 protocol types used. In addition, the route server is a member of the subnet; hence, it also acquires a Layer 3 address. When legacy bridges and ATM-attached end systems are initialized on an MPOA network, they will need to be given information about which protocol they should be using (e.g., IP or IPX), and what end system is providing the function of intra-subnet coordination. In addition, the ATM address of the route server and the default forwarder are required.
Registration, discovery, and routing	The registration and discovery activity occurs just after initialization and concerns the set of exchanges that is used by FGs to inform each other of their existence, capabilities, and domains. Discovery describes reporting edge-attached legacy devices to the subnetwork coordination function. As ATM-attached end systems join the network, they register themselves with the subnetwork coordination function servers that they learned about during startup. When an ATM-attached end system registers, it provides Layer 3 protocol addresses that it supports and places these addresses into the ICFG database. At this point the clients have acquired address and group information and have registered with their servers; they can begin to communicate among themselves.
Data exchange	Data flow through an MPOA system can be either of two types: *unicast* or *multicast*. Unicast can be further subdivided as *default flows* or *short-cut flows*.

The default flow mechanism is similar to a LAN BUS and is used whenever short-cut flows are not possible. Default end system-to-end system communication is accomplished using one of the default forwarding servers (either the RFFG or the DFFG). The use of the RFFG is intended for remote end systems; otherwise, the data is sent to the DFFG. All EDFGs must use the LANE |

Table 5.7 *(Continued)*

Step	Function
	BUS default forwarder. When a packet arrives at a DFFG, it checks to see if the destination ATM end system is one that is registered to that DFFG. If it is registered, then the data is forwarded to the ATM end system. If the address has not been registered, the packet is forwarded using the routing system (e.g., the RFFG). If data arrives at a DFFG and the destination is registered as an EDFG, the data is forwarded to the destination using LANE.
	In order for short-cut data transfer to work, the ICFG and the RSFG must interoperate very closely. Short-cut data transfer is accomplished by establishing a virtual circuit across the ATM network and communicating directly with the ATM destination. After the ARP process is completed, short cut completely does away with intermediate default forwarding systems, providing lower latency and higher quality of service.
Routing protocol support	The MPOA route server is responsible for executing traditional routing protocols with end systems on the legacy media that are reachable through EDFGs or with directly connected ATM end systems. The route server determines which subnets it is responsible for via information flow between itself and the ICFGs.
	When a source attempts to locate another computer in the MPOA system, it consults its local ICFG. The ICFG then forwards the request, on behalf of the source, to the route server (RFFG). If the destination is in another IASG handled by the RSFG, it will generate a response to the ICFG. The ICFG, in turn, forwards the response to the source, which can then establish a VCC. If the RSFG believes that the destination is reachable via another RSFG, it forwards the query to the other RSFG. This pattern is followed until a RSFG is reached that contains the source's IASG. When the RSFG responds to the query, it saves the ATM address of the requester so that if there are any changes in routing or reachability, the source can be notified. When changes occur, the source is forced to disconnect the VCC between itself and the destination and reestablish a new VCC after repeating the ARP.

SOURCE: Based on information from References [2] and [3].

define the flow: In this implementation, it is assumed that the criteria for flow detection is the transmission of more than 5 packets in 1 s.[6]

At the core of an MPOA edge device is an implementation of a bridge (e.g., an 802.1D-compliant bridge with VLAN span). The 802.1D spanning tree finite-state machine controls which ports are blocked and which ports are receiving and transmitting data. All MPOA traffic is subject to the spanning tree control. Upon receiving a PDU, the bridging system performs the source-learning and blocked-port checking. The PDU is then analyzed to find the internetworking protocol. It then uses the combination of ingress port, protocol, and destination MAC to find a bridging table entry. The bridging table gives an egress port and an internetworking flows table address. If MPOA is not interested in the MAC address, the flows table pointer will be null.

The MPOA EDFG function can take action on certain "error" conditions. The egress port is checked for two conditions. If the egress port is blocked, or is the same as the ingress port, then an error indication is returned to the calling entity. If the caller was normal bridging, then the error will be ignored and the PDU will be discarded.

If the flow entry is nonnull, then the destination of the PDU must be checked. The protocol classification has provided enough information to locate the destination. The flows table consists of the following entries:

- Control address for queries
- One entry per detected internetworking address
- Internetworking address
- Occurrence count
- Timestamp
- MPOA ATM address (or null if not yet known)
- Flags (query-sent, VC-in-progress, do-no-VC, and MPOA-tagging)
- VPI/VCI (or 0 if not yet established)
- MPOA tag

---

[6] There are several implementation variations on this aspect of MPOA. The timestamp may be replaced with a more complex structure for detecting flows. The upper-level protocol and port numbers in the PDU may be examined for policy-based decisions to establish a VC even before PDUs have to be seen to warrant a flow. For example, an FTP connection can be recognized and may warrant a VC.

If no entry is found in the flow table for the destination address, then one is created, with an occurrence count of 1 and a timestamp equal to the current time.

If an entry is found, then the occurrence count is updated. If the entry does not have an ATM address, then the occurrence count and timestamp are used to decide if a query is needed. If so, then unless the do-no-VC flag is set, one is built and sent to the MPOA control address in the table. The query-sent bit is used to prevent excessive queries. Meanwhile, the PDU is handed back to the bridging service to be sent. The do-no-VC flag is set to reflect a failure of a query and thereby prevent excessive queries.

If there is an ATM address but no VC (and no VC in progress), then the occurrence and timestamp are used to decide if one is needed. If so, VC establishment is begun, and the VC-in-progress bit is set to reflect this. In the absence of the VC, even if one is in progress, the PDU is handed back to the bridging service for transmission.

If there is an MPOA VC, then the MAC header is stripped from the frame and an MPOA LLC/SNAP header is added. If tagging was provided by the far end, then the tag is also added. The PDU is then sent on the MPOA cut-through VC.

When the bridge PDU handling function detects a flow, an NHRP query is generated. This is sent to the control address associated with the table entry. A query is also generated if an NHRP trigger message is sent by the RSFG or RFFG because it has noticed something that it thinks the EDFG should treat as a flow. If such a message is received, it will have the MAC address in it and it will be received over a control VC. Using this, a cache entry is either created or updated, and the appropriate NHRP query is generated. When an NHRP response is received, it causes an update of the cache entry. The extension fields in the query contain enough information to find the flow table entry. If the query succeeds, the table entry is updated with the ATM address, and a VC is established. If the query results in failure, the do-no-VC bit is set to prevent excessive retries.

The selection of attributes for the VC is a local matter. Usually, UBR (or ABR with 0 minimum cell rate) is used, but other variations are possible.[7] If one chooses to establish VC with varying QoS attributes, it is

[7] As in LANE, MPOA now only supports UBR; it is expected that later releases will support other service classes.

even possible to have multiple VCs, although a more complex table structure is then necessary. Along another dimension, the failure handling may be more complex than has been described here. For example, it is possible to establish time intervals after which failures will be retried.

The EDFG will receive MPOA NHRP requests. These contain egress cache information and may contain a tagging extension. The egress cache information specifies the handling for a received internetworking frame. It specifies a particular protocol, ATM source address, and internetworking destination address. For internetworking PDUs received with that combination (over a cut-through NHRP VC), the egress cache information indications what MAC header should be put on the frame and how the arrival should be treated. MPOA treats all arriving internetworking PDUs as if they had logically arrived over a specific ELAN.

The tagging extension allows the EDGF to attempt to optimize received PDU processing. If the tagging extension is present, the EDFG may provide a tag value that the remote sender will include in all PDUs to the given IP destination.

The egress cache imposition information is stored so that when VCs are established or received from the remote party, the information, including the identity of the RSFG that sent the imposition, is associated with each VC.

One fundamental part of the system is the existence of the flow table and the ATM control address associated with that. Whenever an unknown destination is being flooded onto ATM, a LANE proxy (bridge) performs an LE-ARP operation. If this results in the system resolving the address, an ATM address of a LANE peer is returned. If that LANE peer is also an MPOA device, the ATM device type and ATM control address are returned in an extension to the LE-ARP message. (The EDFG's type and control address are provided in the LE-ARP request.) If the type is anything other than an EDGF, then the EDGF creates a flow entry for the associated MAC address and stores the associated control address in the table. The MPOA type and control address are also stored in information about the LANE VCs that are being established. If the VC is not associated with MPOA, it is stored to prevent repeated inquiries.

LANE clients are permitted to learn about reachability from incoming LANE frames. If the frame is coming in over a VC that already has the LE-ARP tagging information, then that is associated with the learned MAC address. If there is no such information, then this is a remotely initiated VC for LANE. Therefore, using the source MAC of the received frame, an LE-ARP is done in order to get the MPOA control information. When a PDU is received whose LLC/SNAP information indicates that it is MPOA data, then it is processed by the EDFG.

If the PDU contains an MPOA tag (there is a special LLC/SNAP in support of this), then the tag is used to look up to check or find the egress cache information. If tagging is used, the cache entries must be validated. The ATM address in the cache entry is checked against the destination in the PDU. If there is no tag in the PDU, then the arriving VC, protocol, and destination IP address are used to find an egress cache entry. If the entry indicates invalid information, then an error should be sent to the control MPOA device.

If a proper egress cache entry exists, then the MAC header is added to the frame, and the frame is given to the bridging system. The ostensible source of the frame is the port to the ELAN indicated in the cache entry. If the bridging system returns an error, then an error indication is sent to the MPOA control address associated with the cache entry.

Several timers are required in MPOA; most are coarse enough in granularity that a periodic scan of the tables is sufficient. All NHRP-derived entries have a lifetime. Information derived from a local query must be reverified prior to lifetime expiration. Egress information must be discarded if it is not refreshed before a lifetime expires. A periodic updating of the flow-checking table will manage the updating counters for correct direction of flows. Also, this will detect if the flow is not being used, so that the VC may be taken down.

In summary, edge devices are stripped-down routers: Initially they query the RSFG for route information; thereafter, they keep a cache. Individual hosts register themselves to the RSFG when they come up. Edge devices and ATM-attached end systems communicate with the MPOA servers to perform address resolution in a manner similar to the way that it is done in LANE. However, MPOA supports direct mapping of network layer addresses to ATM addresses. This allows for the realization of VLANs based on Layer 3 parameters, and abrogates the need to map Layer 2 VLAN and Layer 3 subnets as in LANE. Because Layer 3 addresses are mapped to ATM addresses, boundaries on Layer 3 subnets are also eliminated.

MPOA, LANE, and CIOA Differences

The basic differences between MPOA and LANE are as follows:

- LANE hides ATM features; MPOA exposes them.
- LANE hides QoS features; MPOA exposes them.
- LANE operates at OSI Layer 2; hence, it is bridging.
- LANE requires no modification to end system protocol stacks.
- MPOA is an evolution of the LANE model.
- MPOA operates at both OSI Layer 2 and Layer 3; hence, it is both bridging and routing.
- MPOA requires modification to end system protocol stacks.

Both LANE and CIOA suffer from the limitation that end systems in different logical subnets, but attached to the same ATM network, must communicate with each other through routers. So, LANE and CIOA are still constrained by the traditional mind-set of segregating networks into subnetworks, and then interconnecting them by routers; for example, as seen in Figure 5.7 for CIOA and in Figures 5.8 and 5.9 for LANE. While this approach was the logical step forward when moving up from bridged internets, and has worked well for a time, it may soon be eclipsed by technologies that allow clients to communicate directly across administrative domains without using routers. Single-hop communication means that end systems will see reductions in processing delays, bottlenecks, and performance degradation introduced by interconnected routers and multiple hops. Jitter is introduced by multiple hops—this is very detrimental to multimedia, digital video, and distance-learning applications.

A limitation of CIOA is not only that it is limited to IP, but that every network-attached device must be ATM-based—in LANE, some of the network-attached devices can be legacy-LAN-(Ethernet or Token Ring) based. MPOA is similar to CIOA in that it supports direct association of IP addresses with ATM addresses. However, the MPOA model is superior because MPOA clients are empowered with the knowledge of the ATM addresses belonging to devices outside of their subnet. Furthermore,

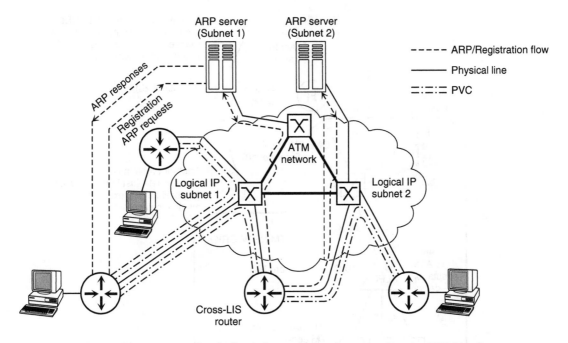

NOTE: The Cross-LIS router implies that the flow is frame-to-cell-to-frame-to-cell—using MPOA implies a more consistent PDU format.

Figure 5.7 *Subnet-based communications in CIOA.*

MPOA provides mechanisms for broadcast and multicast support that are lacking in CIOA.

The advantages of MPOA include the following:

- End systems can establish direct connections to remote servers without using routers.
- End systems experience lower latency in establishing connections.
- The network experiences reduced amounts of broadcast traffic.
- Many of the intrinsic benefits of ATM can be more easily secured.

Transitioning to MPOA

While waiting for MPOA equipment to become available on a broad basis, an organization's choice of LANE versus CIOA is based on the type

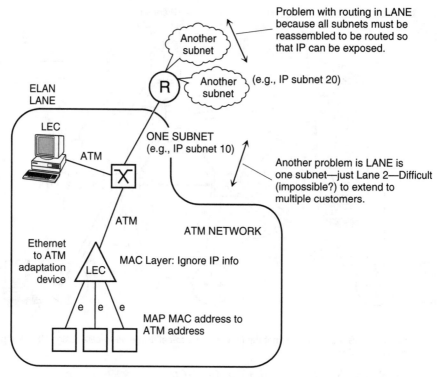

Figure 5.8 *Some limitations of LANE.*

of traffic carried. When performance is important, CIOA should be used; however, if more than one network layer protocol is being used, then LANE may be the short-term solution. LANE enables the network designer to treat ATM as a bridged legacy LAN, while CIOA treats ATM as a point-to-point link between end systems. Furthermore, LANE is also the only ATM-based service that is currently capable of supporting multicast traffic. LANE 2.0[8] allows multiple LESs and LECSs. It also supports RFC 1483 SNAP/LLC encapsulation (VC muxing being used in LANE 1.0). LANE 2.0 is not backward compatible with LANE 1.0. CIOA is simpler to implement, compared to deploying LANE.

Some of the benefits of MPOA include the following:

---

[8] The specification was issued in 1997.

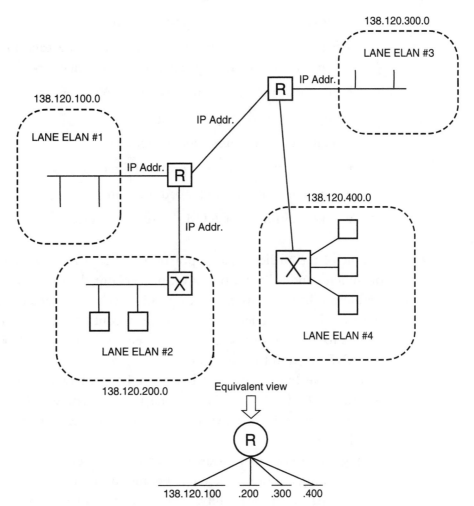

Figure 5.9 *Use of routers in LANE.*

- Achieves better end-to-end performance
- Allows integration with LANE
- Allows MPOA devices to establish direct ATM connections
- Provides a unified approach to Layer 3 protocols over ATM
- Provides for the support of legacy physical layers and legacy rout-
  ing protocols to ensure interoperability with preexisting networks

- Provides the connectivity of a fully routed environment
- Separates switching from routing, optimizing each in appropriate hardware and in appropriate places in the network
- Supports automatic configuration of ATM end systems
- Supports cut-through routing
- Supports direct interdomain connection
- Supports firewalling and protocol filtering
- Supports multicast and broadcast traffic
- Takes advantage of ATM's features and QoS

One of the challenges a corporate network planner faces is to identify, document, and follow a transition plan. Naturally, such a transition plan should not have to involve every possible technological alternative under the sun. For example, a transition plan could be to upgrade existing routers to MPOA; this would obviate the need to go to LANE 1.0, then to LANE 2.0, and then to MPOA.

Today we build networks based on IP subnetworks—LANE works within a flat MAC space (unless one uses routers, which can be slow). Hence, given today's networks, there is a preference to deploy MPOA over LANE.

Single-hop communication between end systems in different subnets without using a router is desirable because of the potential improvements in performance and scalability. This is called *cut-through* routing because it bypasses routers and cuts a path through the ATM network. Cut-through uses route servers to acquire information about the end system on the network, but then relies on Layer 2 technology to communicate with the remote end system. NHRP relaxes the forwarding restrictions of the LIS design (see Figure 5.10).

MPOA affords the ability to support global VLANs that operate independently and beyond traditional router boundaries. End systems that are members of different subnets can communicate directly across the ATM network without passing PDUs through an intermediate, performance-impacting packet-based bridge or router.

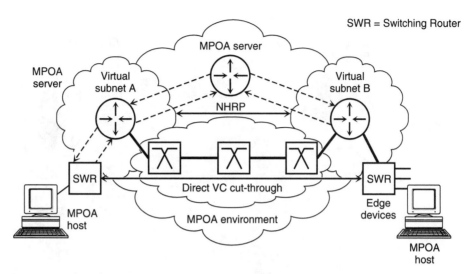

Figure 5.10 Advantages of MPOA in terms of cut-through communication.

Equipment supporting MPOA is now appearing. Figure 5.11 provides one example from Newbridge Networks, an early entrant. In deploying MPOA, the network administrator will need to understand the function and availability of various servers and work with vendors to acquire the right system elements. The core elements of the MPOA solution are the ATM switches and the ATM end systems; these can be the same devices that may already be connected to a corporate LANE or CIOA network. Second, the migration to MPOA will require adding the MPOA servers to the network; this could be as simple as upgrading the ATM switch control software. The third hardware component of the solution will be the addition of MPOA edge devices; this may be achievable via software upgrades to LANE bridges or could require an altogether new device.

Figure 5.12 depicts what a turn-of-the-century network would look like. At face value one can question the value of a standalone LANE 1.0-based corporate network if LANE 2.0 solves a number of restrictions of the earlier version, but incompatible islands arise in the process of deploying one or both technologies. In the short run, however, MPOA and LANE networks are similar; the networks start to differ when their sizes

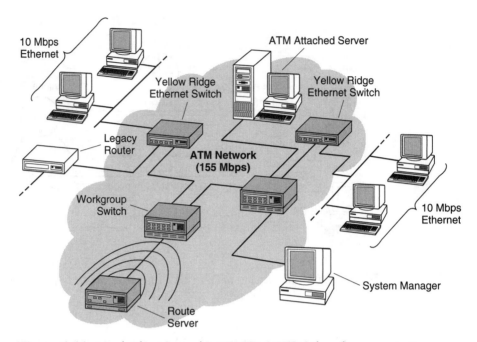

Figure 5.11 *Newbridge Network's VIVID MPOA-based communication.*

grow, or when users require applications that use quality of service. Given that MPOA and LANE are related, one migration approach is to deploy LANE 1.0 with the intent of migrating to MPOA. As the network grows, administrators will be able to create additional virtual workgroups under MPOA that can all share one ATM infrastructure.

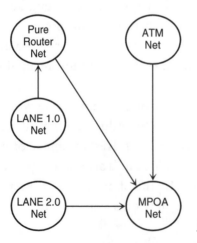

Figure 5.12 *Migration paths to MPOA.*

References

1. J. M. Halpern. "Implementing MPOA in ATM Edge Devices." *Communications System Design* (October 1996): 25–27.

2. D. Minoli and A. Schmidt. *MPOA*. Greenwich, Conn.: Manning, 1997.

3. D. Minoli and A. Schmidt. *Client/Server Over ATM*. Greenwich, Conn.: Manning/Prentice Hall, 1997.

6 *Network Layer Switching Technologies*

The previous chapters have presented detailed discussions of both legacy and innovative switching technologies. And, up to this point, the focus of the vast majority of this text has been on switching protocols that operate at the OSI Layer 2 or the data link layer. Clearly, the ability to switch data-based fast hardware table lookups of the MAC or ATM addresses can produce very fast and reliable networks. However, these technologies also pose critical problems seen by large leading-edge ISPs and in enterprise network design.

Of the problems introduced by large Layer 2 networks, some of the more pressing concerns have to do with the ability of Layer 2 switching to smoothly integrate with Layer 3 switching. There are many vocal opponents of current protocol who claim that the ATM Forum and the IETF have taken radically different approaches. Also, scalability and ease of operation are critical when building large intranets, and this combination is hard to find with a pure Layer 2 switched network. High degrees of scalability can be difficult to achieve with a Layer 2 switched network because the address space is nonhierarchical.

In addition, standards activity and recent network deployments have raised issues surrounding traffic engineering—that is, the ability to use the Layer 2 network to explicitly define the route that data follows. Traffic engineering is one of the more pressing needs, and it continues to drive much of the work described in this chapter. It is a problem that needs to be resolved in order for the Internet to realize continued growth.

This chapter presents protocols developed to facilitate the integration of Layer 2 switching with Layer 3 routing. These protocols produce efficient networking that is capable of being scaled up to very large internetworks possessing both efficiency and solid traffic management capabilities. The information in this chapter is important for network designers and managers because it illustrates the trend in intranet and Internet design, which is toward networks that incorporate hybrid architectures comprised of many cooperating best-of-breed technologies. Network layer switching technologies represent this next-generation class of protocols because, as will be shown, they are made up of the finest points from the data link layer material already covered integrated with robust and proven network layer routing protocols.

Motivations for Network Layer Switching

Before beginning a discussion of network layer switching protocols, it would be helpful to review the pros and cons of data switching at the various layers of the OSI protocol stack. As has been shown in the previous chapters, ATM and other Layer 2 switching technologies have been designed to provide very high speeds coupled with very low latency. This makes them good technological choices for delay-sensitive applications like voice and video data or in cases where high bandwidth is critical, such as ISP backbones.

Layer 3 technologies, such as IP, have been in existence since the early seventies and have, therefore, established their credibility as the dominant protocol for building large-scale, stable networks. Because of the tremendous success of IP, and the lack of strong need for ATM's quality-of-service guaranties, it is difficult to find fault with packet switching—or, more succinctly, it is difficult to find many tasks on the Internet that ATM does better than IP. Some of the most often cited shortcomings of IP networks are associated with either the slow speed or the high price of routers.

Some of the fundamental limitations of IP routers are listed following. These items are directly related to the features that can be provided if the router is combined with an ATM switch.

- Poor scalability to large numbers of physical interfaces per router
- Poor support for very high speed ports—OC12 and above required in enterprise and Internet backbones
- Little support for differing degrees of quality of service and very little experience at implementing quality of service in routers

On the other hand, the fundamental limitations of ATM switches, listed following, are quite different from those of routers.

- Complex signaling and routing protocol
- Little real-world large-scale experience
- Poor integration with Layer 3 technologies (i.e., cumbersome glue provided in the form of MPOA, LANE, Classical IP over ATM, I-PNNI, MARS, etc.)

Even though these protocols each have some serious deficiencies, the promising aspect is that, for the most part, the limitations of Layer 2 and Layer 3 technologies are orthogonal. Therefore, if the two layers could be integrated in a manner such that the coupling was clean, efficient, and realizable, the technology product would be a best-of-breed. The combination that is realized—network layer switching—is a product that combines the IP routing intelligence of a router with the fast data-forwarding skills of an ATM switch.

By combining the strengths of Layer 2 and Layer 3 technologies, network layer switching intends to build a system that has the following capabilities:

- Provides fast and low latency over ATM
- Can redirect data flows or routes onto specially engineered ATM virtual circuits
- Drives the decision to redirect data by an analysis of the data focusing on the application at hand

When the problem is designing new protocols that combine Layer 2 and Layer 3 in a manner that is different than the current IETF and ATM Forum proposals, the body of work can be divided into two categories: flow-based versus topology-based. From a high-level view of network switching, one can think of the *flow-based* models as building a network out of routers or switching devices in which unique ATM virtual circuits are created for each IP *conversion,* where *conversations* are synonymous with a file transfer or WWW session. In the *topology-based* proposals, the routers or switching devices use their ATM fabrics to create ATM virtual circuits that can carry all of the traffic destined between pairs of subnetworks or IP routes.

Most of the work done in defining these new protocols has been contributed by a few major networking vendors. At the time of this writing, there are four major protocol architecture contributions. The flow-based and topology-based contribution categories are on the standards track, primarily within the IETF. In order to keep the material in this chapter manageable and of reasonable length, only the leading proposal from each category that is farthest along the standards process is described in detail.

In the short term, it is generally believed that only the flow-based technology will be available. The reasons for this, as described in greater detail in the following, are its head start in the market and its widespread industry acceptance. The next section of this chapter addresses flow-based network layer switching. There is a great deal of controversy over the use of network layer switching paradigms and a great deal of study has gone into the topology-based models. The general consensus is that flow-based protocol will be used in the LAN and topology-based in the WAN. For this reason, the final section of this chapter describes a leading topology-based proposal, Cisco System's Tag Switching, in detail.[1] Ultimately, both the flow and topology proposals will be merged into a standard that incorporates traits from each.

Network Layer Switching Models

Integration of Layer 2 and Layer 3 means that ATM switching is used as the physical interconnection media and as a high-speed forwarding technology (see Figure 6.1). IP routing technology is then used to propagate routing tables and to determine what path data should follow. If a network designer accepts the preceding arguments, then a final point to be considered is

---

[1] Cisco can be reached at www.cisco.com.

exactly when ATM switching should be employed. There are two different philosophies of how ATM should be used when coupled with IP routers.

The major distinction between the flow-driven versus the topology-driven proposals is the level of granularity applied when considering how IP traffic should be carried over ATM; in some cases these proposals are called *fine-grain* versus *coarse-grain* switching. Fine-grain applies to the flow-based models and coarse applies to the topology-based proposals. However, regardless of the name used, the principal is the same in that they both deal with different abstractions of traffic passing through the network.

When flow-driven IP switching is used to carry user datagrams, the interconnecting IP switches initially employ predefined virtual circuits to interconnect routers carrying the IP default route data flows. In both the flow-driven model and the topology model, network layer routing intelligence is integrated with the ATM switch control software. Therefore, each ATM switch has the ability to understand the traffic crossing the ATM switch from a Layer 3 perspective and consequently to make decisions on how that traffic can be switched as it crosses the ATM network. The IP packet analysis code and the predefined virtual circuits are, in theory, only used for short periods of time. When the routers or switches detect a traffic pattern that triggers the creation of a separate virtual circuit

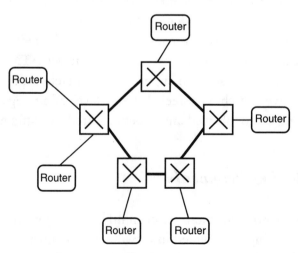

Figure 6.1 High-level view of topology switching.

for the datagrams, the IP control software on the router or switch is no longer is needed to route packets, and the ATM hardware's VCI table is used to forward the cells.

The concept of a router coupled with a switch being able to examine all of the traffic is actually true in both fine- and coarse-grain aggregation; however, in the fine-grain model the router's interaction may be more common. The reason for this interaction is that the switch must be aware of each individual IP session that is crossing the switching fabric. IP sessions can be defined by several different metrics; however, the most common are combinations like the IP source and destination addresses coupled with a TCP port number. The port number will correspond to e-mail, file transfer, or WWW page transfer. Therefore, when the IP switch recognizes a series of packets with the same addresses and port number, it will consider them as a singular flow and will create a segregated virtual circuit to carry the subsequent data.

An alternate method for coupling IP routers with ATM switching is by using topology-based switching, sometimes called *coarse-grain aggregation*. With this method the IP router or switch still uses IP's routing protocols to determine where data should be forwarded; however, the major difference is that when traffic is segregated onto new ATM virtual circuits, it is not done on a per-flow basis. With coarse-grain aggregation, the data patterns recognized are at the level of IP routes. Therefore, each IP switching router at the edge of an ATM network associates ATM virtual circuits with the other routers at the edge of the network.

The arguments for this approach are that the benefits of fine-grain aggregation can be achieved; however, because fewer ATM resources are needed, the system can scale to support more attached devices. In addition, some network architects feel that this level of aggregation matches existing routing protocols well and is better suited to traffic engineering.

Network Layer Switching Proposals

As previously mentioned, there are currently four major proposals for network layer switching, subdivided into two main categories. These proposals are all under review within the IETF, and several have been released as informational RFCs. Informational RFCs are designed only to assist with information dissemination and do not represent IETF standards; however, they can be used to start the process of technological innovation. Ultimately, these various proposals will be studied and the official standard from the IETF will be presented in the Multiprotocol Label Switching (MPLS) RFC [1].

Because the IETF's work in still in its infancy, the information provided in this chapter is somewhat broad and may change as the work progresses. What is clear from the MPLS baseline document is that the IETF's tendency is more toward alignment with the topology-based proposals. However, the vast majority of actual implementations of network layer switching to date are focused on the flow-based models, primarily because they were the first to be introduced and documented in RFCs. Therefore, the material in the next section focuses on the most prevalent flow-based proposal on the market, Ipsilon's IP Switching.[2] The material following IP Switching presents a topology-based network layer switching technology that is closely aligned with the IETF's MPLS baseline—Cisco System's Tag Switching.

Table 6.1 provides a high-level description of the four key proposals in network layer switching. The first two, presented by Ipsilon and Toshiba, are documented in informational RFCs. The proposals from Cisco Systems and IBM are the major contributors to the baseline of the MLPS standard.

Flow-Based Switching: IP Switching

This section discusses the background and operation of IP switching.

Motivations for IP Switching

During the development of the MPOA protocol, some vendors began to consider the body of work generated by the ATM Forum as becoming a

[2] Ipsilon can be reached at www.ipsilon.com.

Table 6.1 Comparison of Network Layer Proposals

	IP SWITCHING	CELL SWITCHING	TAG SWITCHING	AGGREGATE ROUTE-BASED IP SWITCHING
Vendor	Ipsilon	Toshiba	Cisco	IBM
Support for Layer 2	ATM	ATM and Frame Relay	ATM, Frame Relay, and Ethernet	ATM and Frame Relay
Support for Layer 3	IPv4 and IPv6	IPv4 and IPv6	Protocol Independent	Protocol Independent
Interswitch control protocol	Ipsilon Flow Management Protocol (IFMP) [2]	Flow Attribute Notification Protocol (FANP) [3]	Tag Distribution Protocol [4]	Aggregate Route-Based IP Switching (ARIS) [5]
VC distribution or control algorithm	Downstream toward sender	Downstream toward sender	Traditional routing distribution	Downstream multicast
Aggregation model	Traffic-based with per-flow cut-through	Traffic-based with per-flow cut-through	Topology-based per destination prefix routing table entry using routing hierarchy	Topology based on egress ID; each egress has a multicast that sinks flows

large and a difficult software engineering problem. There was a sense that the ATM protocols had become too complex to be practical or implementable because of the number of components that were becoming standards and, subsequently, the number of these that were required to build basic ATM networks. Network architects and vendors felt that in order to support multiprotocol networks using ATM Forum technology, a product would need to support the majority of the following protocols:

- UNI 3.1 for signaling
- P-NNI for routing
- NHRP for short-cut resolution
- LANE for address resolution

- MARS for multicast
- ABR for traffic management

When totaled, these protocols become a very large software systems design process, potentially requiring hundreds of thousands of lines of programming code. To address this problem, vendors began working on simplifying and rewriting existing methods and creating new technology to integrate ATM and IP routing. Their intent was to dramatically reduce the number of ATM Forum standards used when designing a complete IP switching over ATM system.

This section examines the technology designed to effectively join the worlds of IP routing and ATM cell switching by utilizing the best features from each. The development of these protocols at the level of individual TCP data flows has taken many different names, but it is typically referred to as *IP switching* [2]. It is important to keep focused on the goals of an IP/ATM switch—similar to an MPOA network, it identifies streams of packets flowing between a source and a destination, then establishes a cut-through ATM virtual circuit and removes the routers from processing the data packet by packet.

The major difference between the techniques is the fact that IP switching uses only a few simple protocols. In addition, one of the most significant differences between IP switching and multiprotocol label switching (MPLS), or tag switching, as described in a following section, is the level of flow aggregation. IP switching and MPOA typically work on individual IP data flows, while MPLS operates on aggregates of many flows between two routers.

The perceived complexity of the ATM Forum standards led to a new approach to using ATM for networks building that involves selecting the best technology from IP routing and ATM switching and then merging them. One of the first vendors to implement these ideas is Ipsilon, whose products combine the intelligence and control of IP routing, which has evolved on the Internet, with the high speeds and capacity of ATM switching hardware.

The fundamental idea behind IP switching is that the process of routing datagrams is a problem that has been the subject of long analysis and that has already been solved; therefore, these protocols should be used and not reinvented. The excellent performance of IP routing technology has been proven in laboratories, and these protocols have been working and tested on the Internet for years. Therefore, the complexity of P-NNI is not necessary because protocols like OSPF and BGP already exist.

On the other hand, ATM brings to the table high-speed switching and low latency. Joining these technologies—IP routing and high-speed switching—yields products that are based on scaleable, robust, and proven technologies and are capable of providing millions of IP packets per second throughput while maintaining full compatibility with existing IP networks, applications, and network management tools.

The operation of virtual circuit creation requests within the IP Switching system is documented in two IETF Informational Requests for Comments: the Generic Switch Management Protocol (GSMP) and the Ipsilon Flow Management Protocol (IFMP) [2].

IP Switching Operation

An IP switch implements the IP protocol stack on ATM hardware, which operates as a high-performance data link layer accelerator. The actual process of IP switching is rather straightforward. The network is built with a group of edge devices surrounding a core of ATM switches. The edge devices and the ATM switches are running traditional IP routing software like OSPF or BGP. They pass data between themselves along default virtual circuits until they detect a flow, at which point a special virtual circuit is created to carry the segregated flow. It is the responsibility of the edge device to detect the flow and issue the command to create the new virtual circuit.

The IP Switching protocols are designed to dynamically shift between default routing using store-and-forward and cut-through switching based on flow patterns seen in the traffic. The flow recognition is much the same as an MPOA edge device. Because IP switches are fundamentally routers running traditional IP routing technology, they integrate easily into existing internetworks. Default routing decisions are based on IP protocols, so IP switches behave like other IP nodes and are naturally interoperable with existing applications and network management tools.

The operation of an IP switching proceeds along the following algorithm:

1. *Initialization*—interface accounting and routing table creation
2. *Neighbor peering*—establishing peering sessions
3. *Default forwarding*—using IP routing information learned from the peering sessions
4. *Short-cut creation*—after detection of data flow, creating a unique virtual circuit to carry the packet stream

At startup, each IP switch sets up a virtual channel on each of its ATM physical links to be used as the default forwarding channel (see Figure 6.2). IP data traffic can enter on these default ports if passed from an upstream host or edge router equipped with an ATM interface supporting IP switching software. As the IP switch receives incoming traffic from the upstream devices, it uses AAL5 to reassemble the data and examines the IP header via IP routing software on a route processor. The ATM switch hardware functions simply as a high-speed extension of the routing software and as a transport media for interconnecting IP switches.

After the route for the packet is determined, the IP switch segments the packet using AAL5 and forwards it to the correct downstream IP switch, where the process is repeated. At this point, the major difference between the IP switch and a traditional router is that the IP switch records information about the packet header and the direction that it is forwarded. In this way, the IP switch is attempting to perform the same function as an MPOA edge device, finding flows of similar traffic.

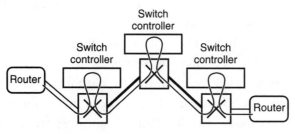

Figure 6.2 *IP switch default behavior.*

From the perspective of the IP switch, a flow is a sequence of IP packets sent from a particular source to a particular destination sharing the same protocol type (such as UDP or TCP), type of service, and other characteristics, as determined by information in the packet header. The IP switch controller identifies flows, since these can be optimized by cut-through switching in the ATM hardware. The part of traffic that is not part of the flow continues to receive the default treatment of hop-by-hop store-and-forward routing.

Once a flow is identified, the switch controller asks the upstream node to segregate the flow using a unique ATM virtual channel. When the upstream node creates the virtual channel, it diverts the traffic to the new virtual channel. Independently, the downstream node can also ask the IP switch controller to set up an outgoing virtual channel for the flow.

When the flow is isolated, the IP switch controller instructs the switch to make the appropriate port mapping in hardware, which will bypass the routing software and its associated processing overhead for all subsequent packets. Therefore, the isolated flow will be switched at ATM speed without the need for hop-by-hop segmentation and reassembly. As time passes, the cut-through path eventually will time out and be deleted when data is not being exchanged over the virtual circuit. After the circuit is deleted, any additional data packets will be rerouted to the default forwarding path. If enough packets arrive to trigger the flow detection code, a new cut-through path will be created and the process will be repeated.

This design allows IP switches to forward packets at rates limited only by the aggregate throughput of the underlying ATM fabric. First-generation cell switching routers support up to 5 million PPS, which is far greater than comparable routers. Further, once the cut-through path is created and becomes operational, there is no need to reassemble ATM cells into IP packets at intermediate IP switches; therefore, throughput remains optimized throughout the IP network.

An advantage of building a system supporting all current major IP protocols is that IP switches integrate smoothly in that environment. For example, QoS requests for each flow are supported using the resource reservation protocol (RSVP), and IP switches can support native IP multicast without any modification to standard multicast protocols, such as the distance vector multicast routing protocol (DVMRP) and the Internet group management protocol (IGMP). Because of this smooth integration, IP switches make an excellent building concentration edge device on a campus network in which the core backbone is pure ATM.

Summary of IP Switching Benefits

There is great debate concerning the applicability of IP switching technology relative to its placement in a LAN or WAN environment. Because IP switches are basically optimized flow detection and cut-through devices, their benefits will be best realized with long-lived IP streams like those seen from the HyperText Transmission Protocol, telnet, or multimedia audio/ video conferencing. In tests they have been shown to perform well in this environment.

An additional benefit of IP switching worth reiterating is that to support the protocol on an existing ATM network requires only upgraded switch control software. IP switching requires no hardware modifications. In addition, several switch vendors have stated support for the flow management protocol and claim that the porting process requires only days to complete.

One potential downside to IP switches is that they may not yield substantial improvements over traditional routers when the IP streams are composed of short-lived protocols such as the simple mail transfer protocol, domain name service, or the simple network management protocol. In these cases the complexity of flow management and virtual circuit creation may outweigh the benefits of cut-through forwarding.

This subsection presents IP switching as described in RFC 1945. To date, this technology has received a great deal of exposure, and it was one of the first to propose the idea of simplifying the union between ATM and IP.

IP switching has received significant industry support in the form of actual implementations of the protocol or statements of intent to support from such vendors as Digital Equipment, Cascade, 3Com, NEC, Hitachi, General DataComm, Ericsson, Fore Systems, Scorpio, and IBM. Therefore, the IP switching flow management protocol should continue to meet with success and realize additional vendor support.

Topology-Based Switching:
Tag Switching

This section discusses the background and operation of Tag switching.

Motivations for Tag Switching

The main motivation for Tag switching[3] [6] is to solve networking problems posed by Internet service providers. However, the technology is not restricted to this environment and can be applied in enterprise networks as well. The problems faced by service providers that Tag switching addresses are as follows:

- Increasing the demands for throughput
- Scaling

> Number of network nodes
>
> Number of routes on routers
>
> Number of concurrent data flows

- Traffic engineering or traffic control—making traffic follow known explicit paths to improve performance
- Improving the ability to incrementally enhance the functionality of IP routing on ISP routers
- Simplifying the integration of ATM and IP technologies

The basic idea of Tag switching is to integrate OSI Layer 3 (network layer, such as IP) control and OSI Layer 2 (data link layer, such as ATM). The advantage of this is to get the benefits of ATM switching technology (i.e., moving data very quickly with optimized hardware) with Internet routing protocols. Therefore, Tag switching attempts to get the best from both technologies. It is important to note that Tag switching support can be implemented with only a software upgrade to routers or ATM switches.

---

[3] Tag switching is currently progressing toward Internet Standard status within the IETF.

Like IP Switching, Tag switching attempts to use only what are thought to be the best aspects of ATM without implementing the majority of the ATM Forum's specifications. For example, Tag switching specifies how routers can use ATM to communicate between ports, but it does not require the following:

- Mapping of IP address or E.164 address
- Mapping RSVP to ATM switched virtual circuits
- Mapping BGP to P-NNI
- Mapping TCP flow control to ABR

An additional benefit of Tag switching that is gained by providing a clear separation between Layer 2 and Layer 3 functionality is the ability of protocol designers to make changes to these technologies' layers without affecting the other layer. There is a subtle distinction that Tag switching provides for integrating these technologies: It tries to improve the inter-working between Layer 2 and Layer 3, while at the same time it provides enough separation to allow changes to either. This is done by dividing Layer 2's forwarding from Layer 3's routing and control capabilities.

Tag Switching Operation

The fundamental building blocks of a Tag switching network are conventional routers and ATM switches, as shown in Figure 6.4. The edge routers are called *Tag edge routers* or *Tag imposition devices*. In the center of the network are ATM switches or more traditional IP routers. Within the large complex network there will typically be a subset of devices that implement the Tag protocol. However, the devices in the center of the network do not necessarily need to support the Tag protocol because they can forward datagrams using pure ATM. Only the edge devices need to support the protocol while the edges mediate between the two protocols.

Logically, the Tag switching protocol is subdivided into two components: forwarding and control. The *forwarding component* is based on label swapping, which is basically the same principal employed by ATM switches as they forward cells. That is, the VPI/VCI conversion from the one received on the input port to the VPI/VCI transmitted on the output

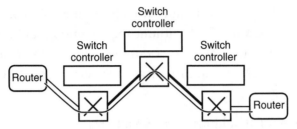

Figure 6.3 *Tag switching architecture.*

port. The *control component* provides the functionality of traditional IP routing. However, it is important to note that the control component can be extended to support protocols beyond IP because the tagging protocol can make tag associations based on *any* Layer 3 protocols. In addition, to help ISPs, the control component can also be extended to support new features for IP routing protocols.

The Tag forwarding component of the system is based on the ability of network devices to acquire the tagging information. In a sense, Tag switching can be viewed as a system where edge routers exchange information to the core devices, as shown in Figure 6.3, so that the core devices pass packets by examining only the VCI (Tag). The core devices, even if capable of understanding IP routing protocols, do not perform routing on a packet-by-packet basis. Like ATM's VPI/VCI, the actual Tag numeric values have significance only on a hop-by-hop basis. The value may change as the Tag path crosses the network. The path created between Tag edge devices typically correspond to a IP route and therefore, provide core grain aggregation.

The core devices receive the command to create interedge router paths via a *Tag distribution protocol* (TDP). This protocol is used to create the core Tag switching device's *Tag forwarding information base* (TFIB). The process is divided into a phase where the routers boot up and begin to exchange routing information along default path or virtual circuits. They next recognize routes that exist between themselves and start a phase of requesting special paths to be created between the routers. These paths are typically ATM virtual circuits.

The routers forward packets down these new paths by placing each data packet inside a special Tag packet. Much like ATM, the intervening switches or routers do not need to examine the contents of the cell or packet to forward the data as it crosses from router to router. One of the major differences between Tag switching and IP switching is that the forwarding decision is based on an aggregation to the level of an IP route as opposed to per-flow switching.

A Tag forwarding information base is made up of the following fields, which should provide all the information required to forward a packet:

- An incoming Tag, which is an index used to quickly find the Tag entry in the table
- An outgoing entry from the table, containing the following:

  1. *Output port*—the port on the Tag switch where the data should be forwarded next
  2. *Output MAC value*—if the output port is a shared media, this value is needed to identify the next-hop router or switch
  3. *Output Tag value* for the next-hop Tag switch

When a packet arrives at a Tag switching router, the input Tag is looked up in the TFIB, the original Tag is removed, the new values from the TFIB are placed in the packet header, and it is then transmitted. In the case of multicast, each incoming entry has multiple outgoing entries.

The Tag distribution protocol updates typically run separate from the routing protocol updates, and the frequency of advertisements can be low. This is because interrouter TDP sessions run over a reliable transport protocol (TCP) and use incremental updates. When the network is being initialized, there will be a large number of Tag update messages; however, once routing is stable, there is no reason to send an additional Tag update unless the topology changes.

Tag updates over a reliable transport also allow routers participating in TDP to quickly realize when sessions have died and, consequently, that routing updates or changes need to be made. There is a subtle point here that is worth clarifying: When using Tag switching without operator intervention, the path the data follows is the same as the IP routing protocol's selected path. Tag switching utilizes routing information at the edges of the network to make its forwarding and path-creation decisions. When IP routing changes, then the Tag forwarding decisions also change.

Within the routers the TFIB will resemble the routing table, with some additional information. This is illustrated in Figure 6.4. The table on an edge router will contain IP address prefixes along with output Tag values. The output table is a Tag advertised by the adjacent router, coupled with the IP prefix. When an IP address packet arrives, the prefix is checked for the *longest match* in the IP routing table. The process of longest match lookup can be considered as part of Tag switching control and need only be performed at the boundaries to the Tag switching network. This overhead processing of locating not just a path to the destination, but the best path from the routing table, is not needed in the Tag switching core.

The best match value is then used to create the next-hop Tag packet. In cases where there are multiple redundant paths, or the TFIB contains more than one entry for some other reason, traditional routing protocol decisions will be used to select the correct path.

The Tag switching behavior previously described is true in the case where an edge or Tag imposition router is transmitting a packet into a Tag-aware infrastructure. When the packet is inside the network, there are two possible means of processing the data flow. First, if the interconnecting device is an ATM switch, the packet is segmented into cells and is subsequently passed down a virtual circuit created with the TDP. Second, if the intermediate device is not an ATM switch but is instead another router that supports Tag switching, then the intermediate router will recognize that it has received a specially formatted packet. It should then perform a lookup in its TFIB for the output Tag and retransmit the packet with the new Tag value.

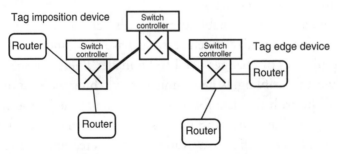

Figure 6.4 Tag switching architecture.

If the Tag switching device is a router, then the Tag is found in a special packet type that has the original packet as the payload (see Figure 6.5). If the Tag switching device is an ATM switch, then the Tag value is just the ATM VCI.

If the interconnecting device is a router, then the Tag table lookup process is very similar to an ATM switch's VC table lookup (see Figure 6.6). Because there is no route computation processing, the packet switching is very fast. It could even be argued that it will be faster to move data with Tag switching routers than with pure ATM switches because data packets are typically larger than ATM cells; therefore, as the VC or TFIB table lookup times become the bottleneck for high-speed switching, the device that switches larger data units will have better performance because it needs to access the table fewer times. Or, conversely, ATM table lookup hardware designs can be relatively easily modified to become higher-performance Tag switching engines.

The frame encapsulation for a Tag packet is shown in Figure 6.7. The encapsulation on frame-based media, regardless of the network layer, is done by prepending a header to the original datagram. The Tag header contains the following:

- *Tag value.*   This is a 20-bit value.

- *Class of services.*   Used by service providers to allow for different grades of services.

- *Bottom of stack.*   Used to signify if multiple levels of Tag encapsulation are being used (i.e., Tag switching inside Tag switching).

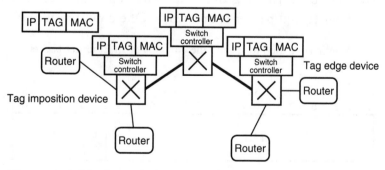

Figure 6.5 *Tag forwarding information base example at edge.*

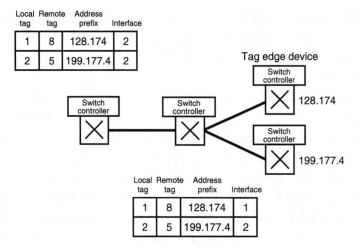

Figure 6.6 *Tag forwarding information base example at edge.*

This bit would signify when the last level of encapsulation has been reached.

- *Time to live.* Because packets follow the same path that IP routing has established, there needs to be a means of detecting that a transient loop exists, that data packets are looping without hope of reaching their destination and should be deleted.

The actual Tag packets have their own formats, which Cisco has registered as new ether types[4] and PPP identifiers. The new ether types allow the interconnecting devices to identify that Ethernet packets do not contain IP headers, for example, but contain Tag switching headers running IP. Therefore, there is a unique ether type for IPX, or any other transport layer protocol, running over Tag switching. The total data header overhead per tagged packet is one word. One common way to look at this is

[4] A designator used to specify the contents of an Ethernet packet.

```
0                   1                   2                   3
0 1 2 3 4 5 6 7 8 9 0 1 2 3 4 5 6 7 8 9 0 1 2 3 4 5 6 7 8 9 0 1
┌─────────────────────────────────────────┬───┬─┬───────────┐
│                   TAG                     │CoS│S│    TTL    │
└─────────────────────────────────────────┴───┴─┴───────────┘
```

Figure 6.7 *Framing.*

that Tag switching over Ethernet networks is done by placing a "shim" between the network and data link layers.

Issues with Tag Switching over ATM

As previously described above, the process of Tag switching has been subdivided into two parts, forwarding and control, providing a clear division of labor. Tag switching can be used over virtually any Layer 2 technology; however, when run over ATM there are some potential pitfalls. The main difficulty with running Tag switching has to do with converting data packets into ATM cells, then correctly mapping the frames onto virtual circuits such that the frames can be correctly reassembled at the destination.

In order to better understand the complexity introduced by ATM, consider Figure 6.8. In the figure, two Tag imposition devices are transmitting data into an ATM network. If the ATM virtual circuits are naively created to correspond to aggregations of routes, then the two inbound flows are merged onto one output flow using the same VCI field. This produces the undesirable effect of creating packets on the output port that are aggregates of the input packet. When the destination receives one of these double-sized packets, it will not be able to reassemble the contents using the specified AAL.

In order to resolve this problem, two possible solutions have been proposed. In the first method, the ATM switch running the Tag distribution protocol realizes that two downstream routers are receiving advertisements for an upstream network. To eliminate the cell-merging problem, the

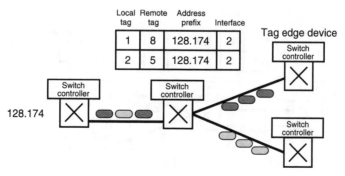

Figure 6.8 *Packet interleaving over ATM.*

switch allocates additional Tags on the output link, one for each down-stream router. That is, the ATM switch assigns a unique Tag per interface to each of the downstream Tag imposition devices (see Figure 6.9).

By creating end-to-end Tag significance, the problem of VCI merge is eliminated in much the same way as end-to-end ATM VCs segregate data flows. However, the Tag distribution protocol is still operating on aggregates corresponding to the abstract level of an IP route. An additional drawback is that the number of Tag imposition device peering sessions will grow from N to N^2.

When an interface-specific Tag is used on the intermediate ATM switch, the cells from the input ports are still interleaved; however, now the output VCI (Tag) values have been changed in their VCI fields. With the unique VCI fields, the cells can then be correctly reassembled at the destination.

One drawback to this solution is the potential reduction of network scale because the number of required Tags per address prefix needs to be increased, and in some topologies the increase can be dramatic.

The second method for coupling with ATM cell interleaving is to force the ATM switch to correct for the problem by transmitting only one frame at a time on the output link. The technique is called *VC Merge*[5] (see Figure 6.10). The switch can perform this function because it can look for AAL5 end-of-message flags to determine packet boundaries. When it receives a complete AAL5 frame it ensures, via buffer management, that it does not interleave the cells. The processing involved with the algorithm

---

[5] VC Merge is the one except to the statement that Tag switching can be implemented with software only. In the case of VC merging, hardware modifications are required.

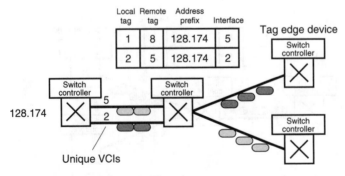

Figure 6.9 *Interface-specific tags.*

is similar to the analysis done by current ATM hardware performing early packet discard, so implementing the feature should not be overly difficult.

There are drawbacks to this technique in that it changes the end-to-end latency performance realized on the VC-merging ATM network to be more like that of a traditional packet-switched network. The latency added is not huge and may be as low as the time required to buffer one packet at OC3c speed. Even if the ATM switch supports a number of port speeds, it will wait until the packet has been completely received into the input buffer before committing to transmitting it on the VC-merged output port. In addition, the data is arriving from a packet network, so introducing a protocol that yields packet network performance seems irrelevant.

The overhead of VC Merge will require hardware modifications; however, this approach will yield a potentially much more scaleable network because the system needs to support only one Tag per prefix or one Tag per exit router. In addition, the ATM switch can run this protocol independent of the VC Merge algorithm. Therefore, the CBR/VBR, and even ATM signaling, can continue to run on the ATM switch supporting Tag switching.

Loop Detection

A final consideration for the protocol's operation is loop prevention. An important consideration for Tag switching is the ability of the Tag distribution protocol to perform loop detection, because this function was once the sole responsibility of Layer 3 protocols; however, the functionality of

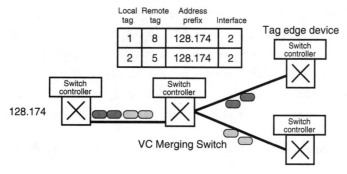

Figure 6.10 *VC merging.*

Layer 2 and Layer 3 have now been coupled, and loops can be created within the ATM network. In order to prevent looping within an ATM infrastructure supporting Tag switching, a potential problem exists if the ATM switches are running a Tag distribution protocol, because this protocol is basically a traditional Layer 3 routing technology that can itself create loops. Once ATM cells begin to be transmitted into a loop there is no way to detect the loop, because ATM switches only look at the VPC/VCI values and do not decrement a time-to-live (TTL) value.

Therefore, in order to prevent loops, the Tag distribution protocol must be able to detect any looping behavior before the Tag forwarding paths are created. In addition, in order for the IP TTL to be correctly set, the Tag imposition device must set the decrement TTL. However, the interconnecting switches will not be able to adjust the value, so in the Tag switching network, the interconnecting fabric will appear as zero hops.

Scaling Tag Switching Networks

A key motivating factor in Tag switching networks is to be able to integrate Layer 2 and Layer 3 technologies so that scaling problems can be reduced. In many traditional IP-routed networks, with several routers surrounding an ATM core, there are potential problems with routing convergence and BGP peering session overloading. These problems can become critical as the number of routers and, hence, the number of adjacencies, increases. The problem is seen in ISP backbones that have been flattened so that the adjacent routers running a link state[6] routing protocol have dozens of peering sessions.

When a group of routers is built with a flattened Layer 2 architecture, each router surrounding the ATM cloud typically peers with every other router. Therefore, every router, from a Layer 3 perspective, is just one hop away from every other router on the cloud. Today's routers and routing protocols can support this arrangement only if the number of attached IP router adjacencies does not exceed 30 to 40. The problem with this net-

---

[6] A routing protocol that determines reachability by examining the operational state of network communication links.

work architecture is that the amount of information transmitted between routers when a link fails is substantial. The problem only gets worse because interrouter data can exponentially increase with the number of attached routers. Essentially, there are two scaling problems that Tag tries to address if link state protocols are used to connect routers across, and there is an ATM core with dozens of router adjacencies:

1. Loading on individual routers to support multiple peering adjacencies (peering sessions)
2. Loading on physical ATM links in support of the increasing number of link state update packets, assuming multicast is not used

Tag switching alleviates this problem because the protocol is supported on the core ATM switches. When Tag switching is running on the ATM switches, it is now an active peer in the link state protocol adjacency relationship. Therefore, the number of peering sessions that the routers around the cloud need to support has gone from every ATM attached router to only one peering session with the Tag switching ATM link (see Figure 6.11).

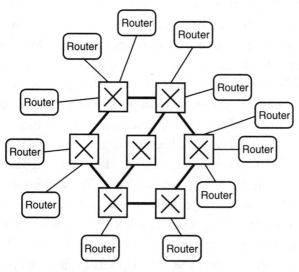

Figure 6.11 *Scaling Tag switching networks.*

Quality of Service in Tag Switching

In a Tag switching environment QoS can be subdivided into two categories, depending again on the level of aggregation. The goal is to create a system where quality of service can be achieved end to end. At a high level of aggregation, Tag switching defines class of service (COS) using the precedence bits of an IPv4 header. This is considered coarse-grain QoS, because the decision to place a higher value on the traffic is made at a router port, or a similar level of abstraction.

In order to map QoS to Tag switching, the Tag imposition device uses the QoS bits of the Tag switching header to mark the packet as higher-value. Therefore, the Tag header instructs the network on where to send the packet and on its value for the QoS. When the packet arrives at the intermediate ATM switch in the Tag network, the switch can examine the QoS bits to determine the scheduling for the packet.

A second, more fine-grain implementation of QoS can be achieved with the resource reservation protocol (RSVP). RSVP is the protocol developed by the IETF as a means to provide fine-grain QoS in the Internet. The RSVP protocol is a signaling protocol that instructs the intermediate devices on what resources will be needed on a per-session basis. For example, RSVP can reserve bandwidth for a single video conference. Once the reservation has been accepted, the routers will identify the higher QoS stream of packets by some unique identifier, like the source and destination IP addresses along with the TCP port number.

When RSVP is used with Tag switching, the result is the same as if the protocol had been run over native routing. The major benefit of Tag switching in this environment is that it is not necessary to map between RSVP signaling messages and ATM signaling messages. This is a major advantage, because the IETF has spent a great deal of effort defining this mapping and there is still concern that it will not work with ATM signaling.

When RSVP is run on Tag switching, each RSVP session becomes associated with a dedicated Tag value. The RSVP source will generate a reservation request message, which eventually reaches the edge Tag imposition device. At that edge router, the RSVP message is converted into a Tag value. This is a very simple mapping of RSVP flow classifications to ATM because the ATM switches process the messages in the same fashion as IP routers. The Tag bindings are carried in the upstream and downstream RSVP reservation messages (i.e., RSVP RESV and PATH messages).

Multicast on Tag Switching

Multicast support can be implemented on Tag switching in a rather straightforward manner. Recall that the Tag forwarding information base carriers an incoming Tag, an outgoing Tag, and an outgoing interface. In the case of multicast, there are several outgoing interfaces. Therefore, the majority of the complexity of multicast support of Tag switching over ATM concerns setting up the forwarding tables.

The basic idea is that Tags are associated with multicast trees and the TDP control messages are used to distribute the Tag bindings associating the Tag values with multicast trees. When a packet arrives on the multicast tree it will be correctly forwarded on the nonTag tree. Also, when it hits the Tag network the imposition router will identify the multicast packet and the correct Tag value for the tagged multicast tree. On shared media, the Tag network uses the link layer multicast capabilities of the media for multicast distribution.

Traffic Engineering with Tag Switching

Traffic engineering is another example of where the separation of control and forwarding in the Tag environment can yield performance improvements. Because the association of Tags or paths across the network can be done in several modules, manual intervention can be used to engineer the flow of data in the network. IP routing software, for the most part, will be responsible for Tag allocation; however, it is just one module, and Tags will be created with the multicast module, the RSVP module, and the traffic engineering module.

In the cast of the traffic engineering module, Tag switching can be used to make traffic go along a different path than the one the IP routing module would have selected. The reason for this manual intervention can be seen from the traffic patterns produced in the networks shown in Figure 6.12.

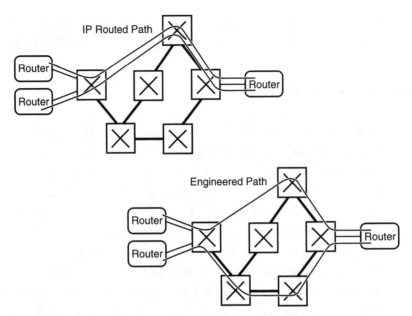

Figure 6.12 *Traffic engineering with Tag switching.*

In the diagram, the network will likely shift traffic along the path with the lowest number of routers along the path. This is because the routing cost of the link is determined by the number of routers passed between the source and the destination. This can result in the undesirable effect of one path becoming congested while the redundant link is underutilized. Clearly, the ideal situation would be to shift traffic from the overutilized links to the quiet paths. Traditionally, this type of traffic engineering is achieved by increasing the link cost for the overutilized links. While this solution may be effective for smaller networks, it often produces unexpected results in large multipath redundant networks.

A better means of traffic control would involve not relying on IP routing to determine the paths used for data flows but, instead, using a manual means. Tag switching allows the network engineer to realize this goal by building end-to-end *tunnels* that mark the path for data across the network. Tunnels correspond to Tag switching paths.

The concept of a tunnel is much the same as that of a dedicated link between two routers. The router at the ingress to the tunnel is responsible for performing the predicated analysis to determine which tag path the current packet should traverse. Therefore, tunnels can be created to carry higher-quality traffic between two subnetworks, or traffic destined to another BGP autonomous system. Before any traffic enters a tunnel interconnecting two routers on a Tag network, it must pass the test of the selected policy criteria.

In a Tag switching network the tunnels are created by humans after analyzing traffic flows. The tools used to create the tunnels can be as simple as link utilization measurements; however, more complex tunnel creation will involve creating tunnels that offer higher QoS and corresponding edge-predicated analysis.

Summary of Tag Switching Benefits

This section describes the benefits of Tag switching. Clearly the major benefit of Tag switching is the ability to better integrate OSI Layer 2 and Layer 3 technologies. This is particularly true with networks where the core switching of the network is done over ATM. With Tag switching the ATM core can simultaneously be used for the multiservice features, such as running CBR and VBR traffic, while it also serves as the basis for the IP routing network. Therefore, network managers should consider building their ATM networks to support several different protocols. Tag switching technology will be used to create efficient mappings of IP over ATM, while at the same time native ATM devices can communicate using ATM Forum–compliant switched virtual circuits.

Tag switching on an ATM core provides several key benefits:

- It provides a very scaleable solution with reduction of routing adjacencies.
- It provides a very good integration of IP protocols to ATM Forum protocols. For example, end-to-end QoS or IP multicast can be transparently implemented.
- It provides traffic engineering abilities that are easily controllable.
- It provides a means to easily scale and control the routing system, because forwarding and routing decisions are separated.

Tag switching provides an interesting mixture of Layer 2 and Layer 3 technologies. It gives a platform that will work on current and future platforms, because it can be realized with a software upgrade. Finally, its separation of forwarding from routing permits a way for network engineers to easily make changes to networks.

Summary

This chapter describes various approaches to network layer switching. Regardless of the exact technical solution, the goals are similar because network layer switching's fundamental motivations are to remove excess computational processing done during packet transmission by dividing routing from forwarding and then remove routing from the process whenever possible. In both the flow-based and the topology-based models, the ATM switch must be aware of IP and capable of participating in IP routing protocols. However, the processes of forwarding and routing still maintain a clear division. Once the control process (i.e., routing) has detected either a route or a flow, it removes itself from the communication path and employs high-speed forwarding from the ATM fabric.

While the various technologies presented in this chapter are mutually exclusive, this is not necessarily the case for Layer 2 switching versus network layer switching in general. Due to the simple economics of most Ethernet-based switching technologies, it seems likely that they will continue to dominate the LAN environment for some time. On the campus backbone the number of choices grows, and the picture is not as clear. However, if the network designer is restricted to the ATM options presented in this and the previous chapter, it seems likely that the enterprise designer should plan on the following architectures.

Switched Ethernet technology will be deployed to the desktop or server environment with the Ethernet hubs interconnected via ATM switches. Within a logical subnetwork interconnected via ATM, the best technology is LAN emulation. The reason for choosing this protocol is that it provides a scaleable means for intra-LAN communication but it also provides a means for the smooth introduction of MPOA. MPOA provides a very good technology for providing inter-ELAN communication and can even be used to route traffic toward ATM-attached servers. MPOA should be able to scale to support networks with several thousand hosts and support high ATM-attached server farm concentrations. As a rule of thumb, flow-driven solutions are best suited for campus LAN and corporate intranets. The only major drawback may be MPOA's lack of support of IP multicast.

For interenterprise traffic or for IP traffic contained within an ISP's backbone, the technology of choice will most likely be topology-based or MPLS. Initially, this takes the form of Tag switching, but once standardized it will be extended to a multivendor environment. MPLA will provide the level of aggregation appropriate for enterprise or ISP backbones (i.e., at the level of IP routes). So, as a rule of thumb, network designers should consider topology-driven solutions for large intranets and the Internet backbones. Of the two models presented, IP switching has the most experience from a functioning implementation standpoint, and while the topology-based proposals lack real-world experience, their IETF backing is critical.

There is no doubt today that the network infrastructure is the most important component of the corporate IT environment. The networking industry has recognized this fact, and the emergence of highly automated management tools is an undeniable sign of this recognition. Consequently, network device manufacturers are introducing new products that already integrate functionality that allows their management through LANs, WANs, and even the Internet. However, an extremely important point to consider is that these management platforms rely on an operational physical layer to provide even the most basic set of their features.

Physical layer switching technology complements current management platforms by providing connectivity management tools that these platforms currently lack. With additional features that directly address troubleshooting and fast intervention requirements, these tools bring a level of control over the physical layer that was not available beforehand. And with increasingly complex network infrastructures, continuous technological migration, and heterogeneous environments, total network control has become the most important item on corporate IT agendas.

The current phase of network layer switching protocol development and implementation is focused on interoperability. For example, the interaction between MPOA and MPLS needs to be better understood and tested. Also, different vendors must prove that their tag distribution or flow management protocols interwork. In addition, RSVP and IP multicast technologies need to receive further attention and implementation experience in enterprise and ISP networks. Nevertheless, from the work that has been completed it is clear that these protocols are both very necessary in large ATM networks and that the complete body of work involved in designing the ultimate MPLS solution is significant.

References

1. R. Callon, P. Doolan, N. Feldman, A. Fredette, G. Swallow, and A. Viswanathan. A Framework for Multiprotocol Label Switching. ftp://ietf .org/internet-drafts/draft-ietf-mpls-framework-00.txt. January 1997.

2. P. Newman, W. L. Edwards, R. Hinden, E. Hoffman, F. Ching Liaw, T. Lyon, and G. Minshall. Request for Comments 1954: Transmission of Flow Labeled IPv4 on ATM Data Links Ipsilon Version 1.0. May 1996.

3. K. Nagami, Y. Katsube, Y. Shobatake, A. Mogi, S. Matsuzawa, T. Jinmei, and H. Esaki. Request for Comments 2129: Toshiba's Flow Attribute Notification Protocol (FANP) Specification. April 1997.

4. P. Doolan, B. Davie, D. Katz, Y. Rekhter, and E. Rosen. Tag Distribution Protocol, Internet Draft. ftp://ietf.org/internet-drafts/draft-doolan-tdp-spec-01.txt. May 1997.

5. N. Feldman and A. Viswanathan. ARIS Specification, Internet Draft. ftp://ietf.org/internet-drafts/draft-feldman-aris-spec-00.txt.   March 1997.

6. Y. Rekhter, B. Davie, D. Katz, E. Rosen, G. Swallow, and D. Farinacci. Tag Switching Architecture—Overview. http://ds.internic.net/internet-drafts/draft-rekhter-tagswitch-arch-00.txt. May 1997.

7 *Physical Layer Switching*

The preceding chapters explore switching technologies that utilize the logical and network layers of the OSI model to direct traffic. As shown, logical layer switches use the MAC address as the primary means for redirecting traffic, while network layer switches use the IP address to accomplish the same goal.

This chapter explores a new emerging technology that utilizes the *physical layer* of the OSI model to provide a path between two points.[1] This method involves manipulating the network cabling and, most important, the connectivity between different network components.

Physical layer switching technology, unlike logical or network layer switching, is not dynamic. At the heart of the concept is a matrix switching architecture that provides the physical path between two points. This path, once established by a command to the matrix, is locked up until one of the two points has to be connected to another. Considering that network services are always at one end of a given path, Physical Layer Switching allows any service to be connected to any port (ASAP).

Because of the electronic nature of its matrix switch, physical layer switching allows MIS managers to control the physical layer right from the desktop. This distinctive feature provides significant management benefits, since network managers now dispose of a tool that allows them to have a set of hands in the telecommunications closet, 24 hours a day.

Most physical layer switches on the market today consist of a hardware matrix switch controlled by software. Typically, the switch is controlled

---

[1] This chapter was contributed by Paul Chalifour, Vice President Technology, CDT/Nodex, Rolling Meadows IL 60008.

from a management station using the SNMP protocol through a dial-up or LAN/WAN link. Figure 7.1 illustrates how a physical layer switch fits into the network. In fact, it can be installed at any point where a change of physical paths is required over time. For example, it can fit between hubs and workstations and between hubs and backbone switches, as well as within the corporate WAN.

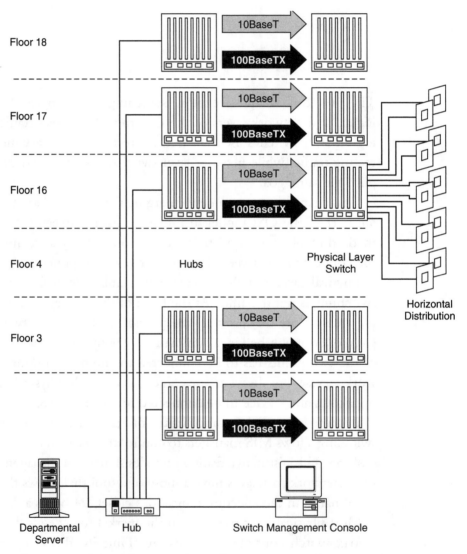

Figure 7.1 *Physical layer switching allows any port to be connected to any service.*

The management software typically allows the user to control the switch from any location and provides a user-friendly interface to facilitate the identification and movement of ports and services. Some manufacturers also incorporate asset management features and reporting into their software—a combination that introduces point-and-click management of Moves, Adds, and Changes.

Importance of Physical Layer Switching

Physical layer switching provides a myriad of benefits to a corporation. Unlike network and logical layer switching, physical layer switching does not require a homogeneous environment in order to provide its functionality. This is an important feature, considering that many LANs today may have up to three networking technologies operating at once. RS-232 lines for printers, Ethernet for office workstations, and 100-Mbps Ethernet for power users—all operating at once—is not an uncommon sight. Physical layer switching provides a cost-effective method for managing these non-homogeneous environments regardless of their size and complexity.

In summary, physical layer switching provides an additional layer of flexibility that increases user uptime and network flexibility while providing service allocation with reduced operational costs.

Increasing Efficiency

One of the primary benefits of physical layer switching is that corporations can increase their operational efficiency regarding the management of the network infrastructure. Most corporations perform basic physical layer switching functionality by dispatching human resources to the telecommunications closet to perform the Moves, Adds, or Changes they need. These resources require time to locate and reach the closet (which can take hours, depending on the office's location), find the proper connection (and many times the connections are badly labeled), and make the appropriate changes. In many cases, this process can take hours, or even days, to perform. As well, bad record keeping (an all-too-common malady) significantly increases the probability that a technician will disconnect a working user, which may result in additional problems to address.

Some corporations attempt to maximize their resources' efficiency by grouping network changes before dispatching the resource to remote sites. This policy, however, causes significant loss of productivity and, potentially, of revenue to the corporation.

A properly located physical layer switch can allow an MIS department to perform Moves, Adds, and Changes in a matter of seconds, as opposed to hours with the traditional methods. This additional efficiency has the side benefit of freeing up resources that can be assigned to higher priority tasks. Furthermore, the automatic update feature ensures that all network changes are accurately logged and registered in the management software's database.

Reducing Operating Costs

The increase in efficiency outlined in the preceding discussion results in a dramatic decrease in operating costs.

It is interesting to note, however, that increasing control over the physical layer can also result in a decrease of operating costs. Due to the traditional methods of executing Moves, Adds, and Changes, patch panels rapidly become congested, which, over time, results in a loss of control. Under these circumstances, technicians could even spend more time trying to find the right service to move, which in turn could result in "short-cuts" that further reduce the flexibility of the patch panel system. Examples of such shortcuts are adding a new patch cord without removing the old and failing to record the appropriate changes to the logbook. Physical layer switching allows corporations to reduce network operation costs through a remotely controlled management of service to user assignment while keeping accurate tracking of the connectivity.

Whether it can be described as an increase in efficiency, a reduction in cost, or an increase in productivity, one of the key benefits of physical layer switching is its ability to manage networks in real time from a central location. Contractors and in-house resources not only represent a significant portion of operating costs, but they also fail to match the ability of physical layer switching to provide immediate troubleshooting of both the logical and physical layers or to respond to individual and group organizational changes.

Increasing End-User Uptime

Contrary to the technologies previously discussed, physical layer switching is the only technology that provides the capability to relocate any LAN resource or user at the click of a mouse. The direct result is an increase in user uptime, which in turn increases productivity by providing the ability to use the physical layer to route around network problems. Consider, for example, that a hub port fails and the resource or user connected to that port is no longer visible on the network. Physical layer switching allows the network administrator to immediately switch the resource to an unused port, restoring network connectivity. This ability to manipulate the physical layer also introduces additional benefits and services, such as service on demand, security (access control), remote troubleshooting, dynamic network configuration, and so forth. These benefits are discussed later in this chapter.

With physical layer switching, the network manager is provided with a tool that allows manipulating the physical layer from a remote location. In fact, this is similar to having an extra set of hands in each telecommunications closet, 24 hours a day, providing a unique opportunity to enhance the network management capabilities that an organization can provide as core or common service and support functions to its end users.

Security

Physical layer switching offers enhanced security in several ways. First, by providing the ability to access telecommunication closets remotely, physical layer switching virtually eliminates the need to access these closets physically. Second, the ability to physically connect and disconnect resources remotely further enhances existing security mechanisms—with physical layer switching, an organization can physically disconnect critical resources or public wall outlets during off hours, adding security to firewalls and similar security mechanisms.

Troubleshooting

Physical layer switching provides the ability to share expensive resources, such as network analyzers and sniffers. From a remote location, for example, an MIS manager can connect network analysis tools to a segment using physical layer switching. This troubleshooting method provides unsurpassed flexibility in locating trouble sources, ultimately resulting in high system availability to end users.

Various physical layer switching manufacturers also provide test cards that allow MIS managers to remotely evaluate and troubleshoot cabling-related problems. These cards are usually controlled from the management software and are able to test the network cabling, rapidly identifying the problem or eliminating the physical layer as the cause of network failure. Considering that a high percentage of LAN problems (up to 50 percent, according to industry analysts) are still attributed to the physical layer, this feature alone is considered to be of great value, especially at remote sites.

Control

Physical layer switching introduces a level of control over the network that is not possible using traditional management methods. Key points that are worth mentioning here include the following:

Online tracking. Most physical layer switching systems have integrated real-time databases that ensure accurate information on user connectivity and related equipment. At any moment, a full assessment of user and equipment connectivity can be conducted.

Responsiveness. Physical layer switching allows for instantaneous network reconfiguration.

Remote management and troubleshooting. Physical layer switching allows troubleshooting and execution of network changes from a central location where highly trained personnel are available.

Improved network availability. Physical layer switching can instantly route around problems, resulting in improved network integrity and reduced network failures. In addition, enhanced diagnostic tools enable network administrators to shorten network downtime to a minimum.

Dynamic Network Configuration

Physical layer switching permits remote Moves, Adds, and Changes. This core functionality enables MIS managers to offer dynamic network configuration for such services as bandwidth on demand, load balancing, and access to limited services. Bandwidth on demand can be implemented either by moving a user to a network segment with lighter traffic load or by connecting the user to a switching hub port or a high-speed port (possible between 10BaseT and 100BaseTX with the appropriate network cards).

Load balancing allows MIS personnel to analyze the load on different segments and use physical layer switching to rearrange the users for optimum traffic flow.

Access to limited resources, such as video conferencing, can be achieved by assigning a segment specifically for this application. At any given time, only users of the video conferencing application will be assigned to this segment of the network, which will result in a full bandwidth and uninterrupted service.

Service on Demand

Remote Moves, Adds, and Changes also allow MIS managers to offer service on demand in telecommuting or office-sharing scenarios, where multiple users share common office space. Multiple types of services could be provided, based on the users' requirements. Furthermore, a pool of fast ports (such as 100BaseTX) could be made available to accommodate users that need high bandwidth for limited periods of time.

Network Reconfiguration

Physical layer switching offers MIS managers the ability to reconfigure the network from the desktop. The requirement for reconfiguration could result from a device failure, an employee location change, an addition of service, a change of type of service, or from maintaining proper LAN segmentation and load balancing. Each of these possibilities are explored in the following:

Load balancing.   Physical layer switching, in conjunction with network management tools, can be used to balance segments directly from the manager's desktop computer, ensuring proper network performance.

Migration.   Physical Layer Switching allows easy migration from one technology to another. Most LAN environments are in a constant state of migration: RS-232 or 3270 point-to-point protocols are giving their places to LAN technology, 10-Mbps Ethernet is moving to 100-Mpbs Ethernet, shared is being migrated to switched, and so forth. This migration activity is not expected to cease, since after the migration to a new platform, a better platform will be installed, which, in turn, will trigger the next generation of migration activity.

Physical layer switching can simplify migration to the next generation of technology by reducing the number of interventions required in the closet. This is accomplished by offering a one-time termination of the new device in the closet. Subsequent interventions to the network can then be executed remotely, allowing the allocation of services when they are required, right from the desktop computer.

Allocation of bandwidth.   Physical layer switching technology allows the MIS department to control the bandwidth allocation to end users without a complete forklift of its current technology. This is accomplished by terminating shared and switched LAN equipment ports to the physical layer switching device and reconfiguring the LAN segment size from the management software. MIS personnel can thus dynamically reconstruct segment sizes or assign full bandwidth to a user with a mere mouse click.

Fallback.   Using the integrated database, physical layer switching technology allows for rapid fallback to the previous configuration when a new implementation encounters deployment problems.

Deployment of Physical Layer Switching

Physical layer switching can be beneficially introduced in any network environment. This is primarily due to the fact that most switching devices are completely transparent to the type of data signals used. This transparency is fundamental in allowing the preceding benefits and the following deployments to happen.

Physical Layer Switching in Shared Environments

Shared network equipment still carries the lion's share of corporate network connectivity. The constant striving for higher bandwidth allocation, better network traffic management, and broadcast storm avoidance has resulted in the implementation of smaller LAN segments, which are either bridged or routed (see Figure 7.2).

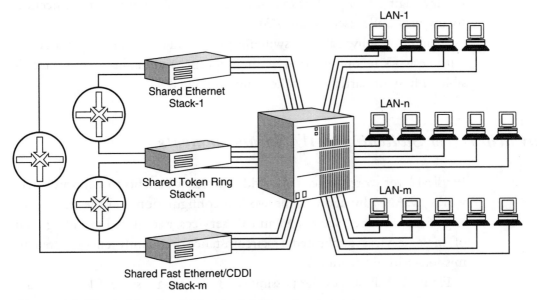

Figure 7.2 *Physical layer switching in shared environments.*

Dynamic workgroups, a highly mobile workforce, and migratory activity characterize current corporate dynamics. The challenge here is to establish a flexible networking infrastructure that can address these dynamics and their associated requirements. These changing requirements may force the implementation of multiple LAN technologies while maintaining the requirements for fast service to the user, high system availability for maximum productivity, and reasonable costs of operation. Physical layer switching features discussed earlier in this chapter provide an efficient and cost-effective method for leveraging current investment in shared networking technology while specifically addressing issues related to dynamic workgroups and technological migration.

Physical Layer Switching in Switched LAN Environments

In either switched or VLAN environments, physical layer switching can add benefits by allowing control over the physical layer. Furthermore, since the cost of equipping every port in the network with switched LAN capability can become very expensive, physical layer switching can permit the leveraging of a pool of expensive ports by allowing their connection to any wall outlet (see Figure 7.3).

In addition, physical layer switching allows all of the benefits associated with network reconfiguration, testing, uptime, and migration to be accessible in fully or partially switched environments.

Physical Layer Switching in WAN Environments

Physical layer switching can also add significant benefits to wide area networks (WANs) by allowing network reconfiguration and the sharing of expensive resources, such as T1 and T3 services, as opposed to having each of these services duplicated (a circumstance frequently encountered in mission-critical situations).

Figure 7.4 illustrates the possibility of assigning a spare T1 line to any router that may have lost service on its primary interface.

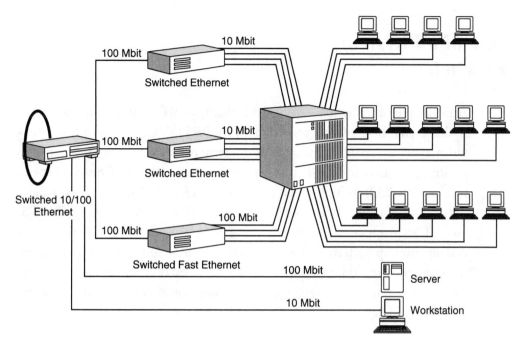

Figure 7.3 *Physical layer switching in switched LAN environments.*

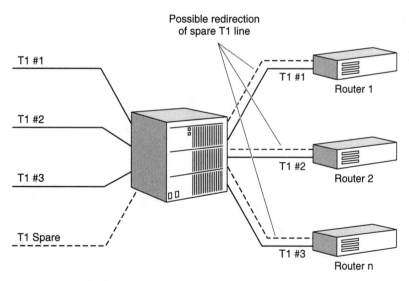

Figure 7.4 *Assigning a spare T1 line to any router.*

Example of a Physical
Layer Switch

An example of a physical layer switch that is available on the market today is the DynaTraX switch provided by NORDX/CDT. DynaTraX provides a switch matrix that can be incremented in 18-port segments to create larger matrices. It is transparent to data protocols from low speed (such as RS-232) to high speed (including ATM 155 Mbps and fast Ethernet) as well as other data protocols (such as 10BaseT, Token Ring, T1, T3, etc.). The product is controlled using SNMP over the network with a dial-up backup system. It has been successfully deployed in many industry segments, including financial and educational applications, laboratories, and systems providers; the users have found the technology to be very beneficial.

8   *Fibre Channel Standard and Technology*

Internetworking of high-performance computers has become the focus of much attention in the data communications industry.[1] Performance improvements in processors and peripherals, along with the move to distributed architectures such as client/server, have spawned increasingly data-intensive and high-speed networking applications, such as medical imaging, multimedia, and scientific visualization. However, the interconnects between these systems and their input/output devices cannot keep up with the blinding data rates, nor can they provide the distances needed for local area networks spanning campuswide areas.

Fibre Channel, a switched protocol allowing concurrent communication among workstations, supercomputers, mainframes, data storage devices, and other peripherals, provides a total network bandwidth on the order of a terabit per second. Fibre Channel is 10 to 250 times faster than existing networks, capable of transmitting at rates exceeding 1 Gbps in both directions simultaneously. It is also able to transport existing protocols, such as IP, SCSI, HIPPI, and IPI, over both optical fiber and copper cable. Fibre Channel products are already appearing on the market, including switches, adapter cards, and chip sets.

As discussed in Chapter 2, Fibre Channel methods are being contemplated for inclusion in the evolving gigabit LANs. Since LANs provide switched services, the concept of FC is of interest not only for its own sake (and the kind of switching it supports in its own context), but also as a supportive technology for LANs.

---

[1] This chapter was supplied by the Fibre Channel Association. It provides a proper balance to this book by describing an alternative broadband technology.

Standards are meant to be exhaustive, legal documents, and do not always make the best reading material. This brief tutorial provides a readable overview of the standard and its applications.

High-Speed Data Communications

New demands in architecture, performance, and implementation are appearing across the entire spectrum of computer systems, from supercomputers to PCs. These requirements are driven by emerging data-intensive applications, such as atmospheric modeling, multimedia, geographic information systems, and medical imaging, which require extremely high data rates. Microprocessors operating at clock rates of 50 MIPS are common today, and soon processors with capabilities of hundreds of MIPS will be commodity items.

According to Amdahl's law, 1 Mbps of input/output (I/O) capability is needed for every MIPS of processor performance. Current communications standards top out at just over 100 Mbps, not nearly fast enough, as technical computing applications already demand processors exceeding 1000 MIPS. The deficiencies in current transmission rates result in the communications channel becoming a bottleneck to system performance.

New interconnection demands are also appearing. The availability of inexpensive, high-performance workstations and intelligent mass storage systems (soon capable of managing terabytes of data) is providing an attractive alternative to the supercomputers historically used for such data-intensive applications. A more cost-effective system can be implemented by clustering a number of workstations, each operating independently, and linking them to mass storage and display subsystems. In addition, desktop workstations often need to access supercomputers from a distance, such as from a nearby building or across a campus. Networks of supercomputers that allow application designers to treat multiple high-performance computing resources as a single system are also emerging (see Figure 8.1).

All of these developments are causing fundamental changes in the way that high-performance computers and peripherals are connected. While

Business Campus

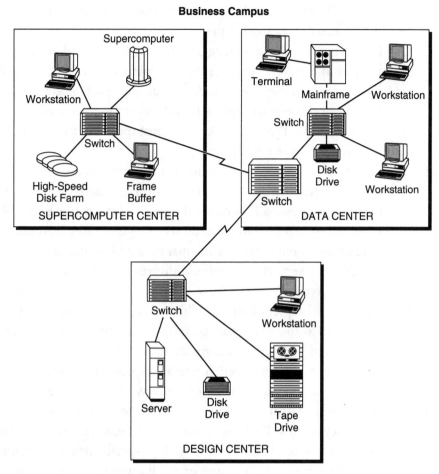

Figure 8.1 *New demands in high-speed data communications.*

computers have become faster and capable of handling larger amounts of data, the network interconnects between computers and I/O devices are unable to run at the speeds needed.

Channels and Networks

The two basic types of data communications connections between processors, and between processor and peripheral, are channels and networks. A *channel* provides a direct or switched point-to-point connection between communicating devices, typically between a processor and a peripheral device. Its primary task is merely to transport data at the highest possible speed with the least delay, performing simple error correction in hardware. In contrast, a *network* is an aggregation of distributed nodes, such as work-stations, file servers, and peripherals, with its own protocol that supports interaction among these nodes. Typically, each node contends for the transmission medium, and each node must be capable of recognizing error conditions on the network and must provide the error management required to recover from them.

A traditional channel, being hardware-intensive, typically has much lower overhead than a network. Conversely, networks tend to have relatively high overhead, because they are software-intensive. Networks are expected to handle a more extensive range of tasks than channels, as they operate in an environment of unanticipated connections, while channels operate in a very clearly defined domain. In the closed system typical of channels, every device address is known to the operating system, either by assignment or predefinition. This configuration knowledge is extremely important to the performance levels of channels. A number of different channel and network standards exist, but they all have limitations regarding the type of high-speed connectivity now required.

A flexible, high-performance, yet low-cost interface is needed that will allow both existing and future systems to communicate at gigabit-per-second data rates. Existing channels and networks are inadequate to meet the mandate of Amdahl's law as high-performance computing systems evolve (see Figure 8.2). Some of the limitations are as follows:

- Distance constraints preclude broadly distributed systems.
- Speed is constraining.
- Connector footprints are too large for shrinking systems and peripherals.
- Physical incompatibility prohibits connectivity.
- Restrictions exist on the number of nodes that can be accommodated.

In 1998, the American National Standards Institute (ANSI) X3T9.3 committee chartered the Fibre Channel working group to develop a practical, inexpenive, yet expandable method for achieving high-speed data

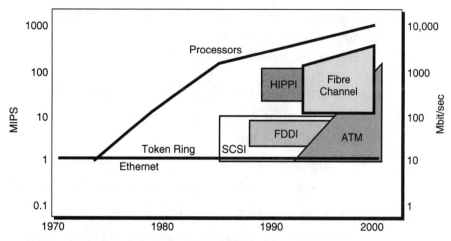

Figure 8.2 *The communications bottleneck.*

transfer among workstations, mainframes, supercomputers, storage devices, and display devices. The Fibre Channel standard addresses the need for very fast transfers of large volumes of information, while at the same time relieving system manufacturers from the burden of supporting the variety of channels and networks currently in place.

Fibre Channel Overview

After a lengthy review of existing equipment and standards, the Fibre Channel working group realized that, since other interfaces being studied (such as the Small Computer System Interface [SCSI] and the High Performance Parallel Interface [HIPPI]) needed similar extension of their next-generation architecture, they should be able to share the same fiber. (Note that *fiber* is used throughout this book as a generic term that can indicate either an optical or a metallic cable.)

Since system integrators frequently support two or more interfaces and software was already in place for all the interfaces that needed to be supported, sharing a port and media was a real possibility. This would reduce hardware costs, both recurring and nonrecurring, and also would reduce the size of the system, since fewer parts would be needed, through multiplexing and shared bandwidth. The new standard was therefore conceived as a common, efficient transport system—a *backbone* underlying and supporting multiple protocols.

The requirements are ambitious:

- Performance from 133 to 1062 mbps on 1 fiber
- Support for distances up to 10 km
- Small connectors
- High-bandwidth utilization with distance insensitivity
- Greater connectivity than existing multidrop channels
- Broad availability (i.e., standard components)
- Support for multiple cost and performance levels, from small systems to supercomputers
- Ability to carry multiple existing interface command sets, including IP, SCSI, IPI, HIPPI-FP, and so on, preserving current driver software

The solution has been to define a channel-network hybrid, containing enough network features to provide the needed connectivity, distance, and protocol multiplexing and enough traditional channel features to retain simplicity, repeatable performance, and guaranteed delivery. This interface is known as the *Fibre Channel*. Documents are currently being developed to use Fibre Channel as a *generic transport mechanism* for command sets of a number of existing interfaces.

Although it is called Fibre *Channel,* the architecture as specified does represent a true channel and network integration. It allows for an active, intelligent interconnection scheme, called a *fabric,* to connect devices. All a Fibre Channel port has to do is to manage a simple point-to-point connection between itself and the fabric. The transmission is isolated from the control protocol, so different topologies—point-to-point links, rings, multidrop buses, and crosspoint switches—can be implemented, depending on the needs of a particular system. The fabric is self-managed—that is, nodes do not need station management functionality, which greatly simplifies implementation.

Fibre Channel is somewhat analogous to today's telephone system. Just as a telephone user can dial a unique "address"—the telephone number—

and be connected to another telephone, a modem, a fax machine, or other device, Fibre Channel makes use of unique addresses to connect processors to other processors or to peripherals, over distances of up to 10 km per link. The fabric functions as the telephone exchange. Where other connectivity standards are limited in the number of possible connections (SCSI, for example, can support only 16 nodes), Fibre Channel allows 16 *million* nodes on a single fabric.

Figure 8.3 shows the existing channel architecture of IPI, SCSI, and HIPPI, compared with the Fibre Channel architecture. A special set of terms that had no prior connotations in the lexicons of the other protocols has been developed for Fibre Channel. The special terminology is shown in Table 8.1. Terms in the Fibre Channel lexicon are capitalized throughout this chapter, and each term is defined as it is encountered in the text. Definitions of these and other terms related to high-speed connectivity are also provided in the glossary at the end of the chapter.

In Fibre Channel, information can flow in both directions simultaneously—data can be transmitted on one fiber and separately received on the other fiber. The structural organization of Fibre Channel supports multiple physical variants at 132.8125, 265.625, 531.25, and 1062.5 Mbps (defined to carry a payload of 12.5, 25, 50, and 100 MBps, respectively). The IBM-patented 8B/10B encode/decode scheme—the same one used in IBM's 200-Mbaud ESCON interconnect system—has been chosen as Fibre Channel's transmission code. The framing protocol supports variable-length frames, security functions, and extended addresses, along with a small built-in command set to provide configuration management and to support error recovery.

Fibre Channel supports both large and small data transfers. It is especially effective in situations where large blocks of data must be transferred within buildings, between buildings, and over campus-range distances. Supercomputers, mainframes, superminis, and workstations can all employ Fibre Channel, and peripherals can include disk drives, tape units, high-bandwidth graphics terminals, high-speed laser printers, and optical mass storage devices.

The high-speed, low-delay connections that can be established using Fibre Channel make it ideal for a variety of data-intensive applications. Many commercial, scientific, and educational environments face the com-

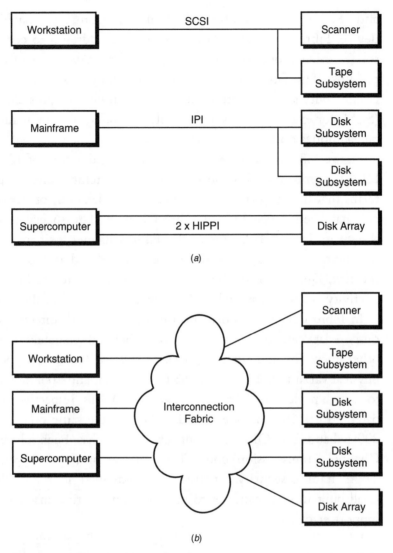

(a)

(b)

Figure 8.3 *Channel architecture: (a) existing channel architecture, and (b) Fibre Channel architecture.*

munications bottleneck resulting from the limited transmission speed of today's interconnect technologies.

Hospitals and medical centers could enable medical staff to perform online consultations simultaneously by transmitting images from magnetic resonance scans or X rays to hundreds of workstations throughout a cam-

Table 8.1 Fibre Channel Lexicon

SCSI	has	Initiators	and	Targets
IPI	has	Masters	and	Slaves
HIPPI	has	Sources	and	Destinations
FC	has	Originators	and	Responders

Words with special meaning:

Connection	Exchange	Sequence
Dedicated	Datagram	Intermix
Login	Multiplex	Frame
Fabric	Port	

pus. Such networks already exist, and workstations have the video capability to display ultra-high-resolution images. But without high-speed communications, real-time video and online consultations are impossible.

Universities, research centers, and large corporations all need to link their various mainframes, minis, workstations, and personal computers to high-speed networks over large campuses. Today, physically distant desktop computers cannot tap into the raw processing capability of a supercomputer or large mainframe, limiting the effectiveness of shared resources in a dispersed environment.

In other environments, users do not need to share computing power, but they do need access to central mass storage in the form of large disk arrays, both magnetic and optical. The insurance industry, which needs to record every single claim and policy form, is rapidly moving to image-based document management systems. Insurers can scan forms directly into the computer and store them as images in optical jukeboxes. Economical, high-speed optical links are needed to provide rapid retrieval of this data-intensive visual information.

High-speed serial communication also permits real-time video conferencing between workstation users, as well as multimedia applications employing voice, music, animation, and video. While all these are currently possible on single computers or in one-room environments, high-speed communications will make them possible over greater distances and, eventually, worldwide.

Fibre Channel can provide high-speed links to massive, intelligent storage systems. Improved transfer rates are especially important in applications like medical imaging, visualization, image-based document systems, geophysical mapping, satellite imaging, or large CAD sets where file sizes can easily grow into the multigigabyte range.

A fast SCSI parallel link from a disk drive to a workstation can transmit data at 160 Mbps, but it is restricted in length, requiring the disk drive to be located no more than a few feet from the computer. In contrast, a quarter-speed Fibre Channel link transmits information at 266 Mbps over a single, compact optical cable pair up to 10 km in length. This allows disk drives to be placed almost anywhere on a campus and enables more flexible site planning.

Another application where Fibre Channel shines is in *mirroring* or *shadowing* data. Mission-critical data, such as information needed for disaster recovery, must be frequently backed up and then stored in a safe location. Typically, this involves downloading to a tape or other portable medium and physically carrying it to a vault. Using Fibre Channel, data can be backed up as frequently as needed directly to a mass storage device at a secure remote location, greatly simplifying the catastrophe plans of any large corporation.

Figure 8.4 shows a typical Fibre Channel environment, with several clusters of workstations, mainframes, a supercomputer, and assorted peripherals in two different buildings within a large computing center. All of these devices need to be connected for high-speed multiuser computing.

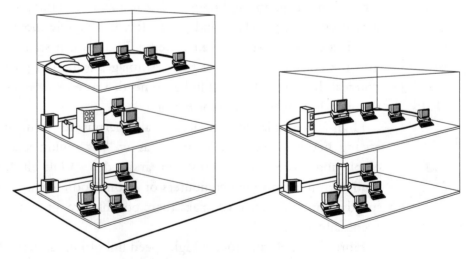

Figure 8.4 A typical Fibre Channel environment.

Physical and Signaling Layer

The Fibre Channel structure is defined as multilayered functional levels. The five layers of Fibre Channel, shown in Figure 8.5, define the physical media and transmission rates, encoding scheme, framing protocol and flow control, common services, and the upper-level protocol interfaces.

In this section, we describe the *Fibre Channel Physical* (FC-PH) standard, which consists of the three lower levels, FC-0, FC-1, and FC-2. The upper levels are discussed in the next section.

The three FC-PH levels are defined as follows:

- *FC-0.* Covers the physical characteristics of the interface and media, including the cables, connectors, drivers (ECL, LEDs, short-wave lasers, and long-wave lasers), transmitters, and receivers.

- *FC-1.* Defines the 8B/10B encoding/decoding and transmission protocol used to integrate the data with the clock information required by serial transmission techniques.

- *FC-2.* Defines the rules for the signaling protocol and describes transfer of the data frame, sequence, and exchanges.

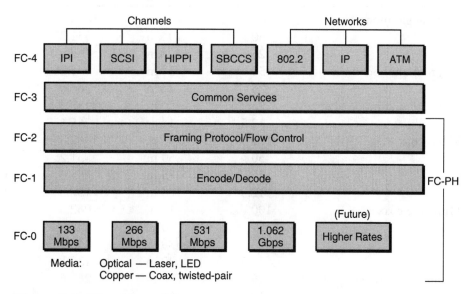

Figure 8.5 *Fibre Channel layers.*

FC-PH completed its first public review in January 1993, and the second review began in October 1993. The three levels are described in detail in the following subsections.

Physical Interface and Media: FC-0

The lowest level, FC-0, specifies the physical link of the channel. The main purpose of Fibre Channel is to have the selected protocol operate over various physical media and datarates, and the physical media described in FC-0 are conservative guidelines for the industry. This approach ensures maximum flexibility, allowing existing cable plants and a number of different technologies to be used to meet a wide variety of system requirements. Table 8.2 illustrates the relationship between the media type and the operating range for each Fibre Channel–defined effective data transfer rate.

Table 8.2 Relationship between Media Type and Operating Range and Speed

TYPE	CONNECTION SPEED, Mbps	MAXIMUM RANGE	LINE SPEED, Mbaud	DRIVER
9-μm single-mode fiber	100	10 km	1062.5	Longwave laser
	50	10 km	531.25	Longwave laser
	25	10 km	265.6	Longwave laser
50-μm multimode fiber	50	1 km	531.25	Shortwave laser
	25	2 km	265.6	Shortwave laser
62.5-μm multimode fiber	25	500 m	132.8	Longwave LED
	12.5	1 km	265.6	Longwave LED
Video coax	100	25 m	1062.5	ECL
	50	50 m	531.25	ECL
	25	75 m	265.6	ECL
	12.5	100 m	132.8	ECL
Miniature coax	100	10 m	1062.5	ECL
	50	20 m	531.25	ECL
	25	30 m	265.6	ECL
	12.5	40 m	132.8	ECL
Shielded twisted–pair	25	50 m	265.6	ECL
	12.5	100 m	132.8	ECL

As an example, with Fibre Channel a single-mode fiber can enter a building, and then a multimode fiber can be used for the vertical distribution inside, with copper drops to individual workstations. For short distances, Fibre Channel permits coaxial connections up to 800 Mbps. On a single fiber or wire, up to 800 Mbps of effective transfer rate is possible with a line speed of 1.0625 Gbaud. A representative FC-0 link is shown in Figure 8.6.

For optical fiber interfaces, Fibre Channel uses a special low-cost duplex SC connector to ensure polarization and ease of use. For the shielded twisted-pair (STP) copper interface, a 9-pin D-type connector is used. For the coaxial connections, a TNC receiver connector and a BNC transmitter connector have been chosen. Each of these connector types is shown in Figure 8.7.

Transmission Protocol: FC-1

The FC-1 level defines the byte synchronization and the encode/decode scheme. Fibre Channel makes use of a DC-balanced 8B/10B code scheme, selected for its superior transmission characteristics. This well-balanced code allows for low-cost component design and provides good

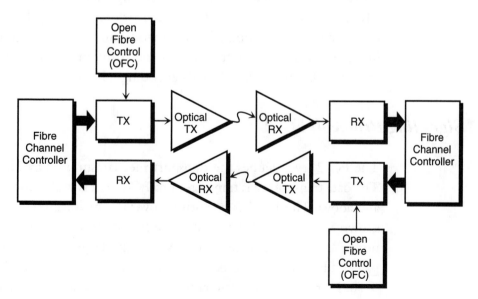

Figure 8.6 *A representative Fibre Channel link.*

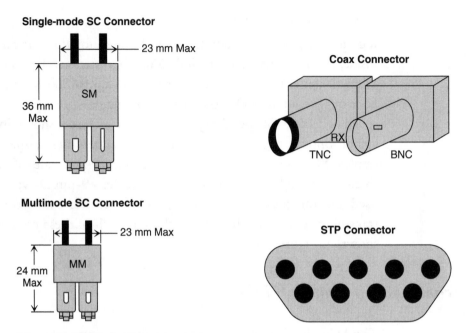

Figure 8.7 *Fibre Channel connectors.*

transition density for easier clock recovery. A unique special character, called a *comma* character, ensures proper byte and word alignment. Other advantages of this code are its useful error detection capability and a simple logic implementation for both encoder and decoder. In this scheme, 8 internal bits (1 byte) are transmitted as a 10-bit group. The block diagram in Figure 8.8 shows the 8B/10B encoding scheme.

Signaling Protocol: FC-2

The FC-2 level defines the transport mechanism used by Fibre Channel. The data transported is transparent to FC-2 and is visible to the FC-3 level and above. The FC-2 is performed over the physical model (FC-PH), as shown in Figure 8.9. Physically, the link consists of a minimum of two nodes, each having at least one N_port (node port) interconnected by a pair of fiber or copper links, one outbound and one inbound.

An N_port is a hardware entity, within a node such as a mainframe or a peripheral, at the node end of the link. Data communications are per-

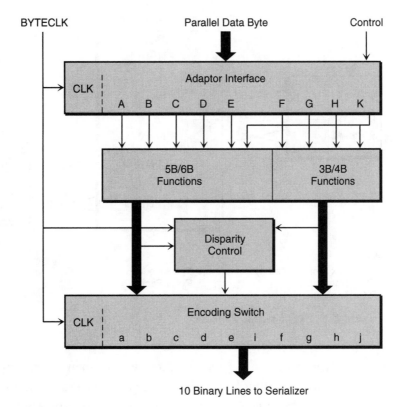

Figure 8.8 *8B/10B encoding.*

formed over the link by the interconnected N_ports. An N_port includes a *Link_Control_Facility* that handles the logical and physical control of the link for each mode of use; it provides the logical interface to the rest of the end system. Essentially, it is part of the termination card that contains hardware and software supporting the Fibre Channel protocol.

Each port can act as an *Originator,* a *Responder,* or both, and contains a transmitter and a receiver. Each port is given a unique name, called an *N_port Identifier.* Simultaneous, symmetrical bidirectional data flow is supported by FC-2.

FC-2 provides a rich set of functions in addition to defining the framing structure. These functions include the following:

- A 32-bit cyclic redundancy check (CRC) to detect transmission errors.

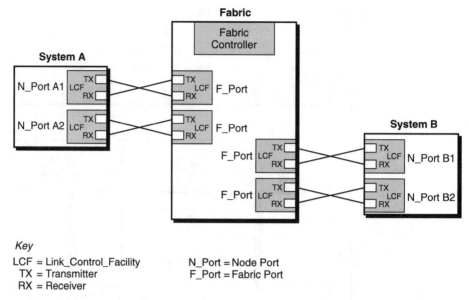

Figure 8.9 *FC-2 physical model.*

- Various *classes of service* to provide and manage circuit switching, frame switching, and the datagram services through the fabric. These include *Class 1,* a hard- or circuit-switched connection; *Class 2,* a connectionless, frame-switched link providing guaranteed delivery and receipt confirmation; *Class 3,* a one-to-many connectionless service with no confirmation of receipt; and *Intermix,* which provides some of the functionality of both Class 1 and Class 2. These classes are reviewed in detail in the following sections.

- Constructs to efficiently support multiplexing of operations.

- A look-ahead, sliding-window flow-control scheme that also provides a guaranteed delivery capability. This flow control differs depending on the class of service being used.

- A set of generic functions that are common across multiple upper-level protocols.

- A built-in protocol to aid in managing the operation of the link, to control the Fibre Channel configuration, to perform error recovery, and to recover link and port status information.

Hardware Control

Fibre Channel is hardware-intensive and relies on the contents of the frame header to trigger actions such as routing arriving data to the correct buffer. That is, commands and responses are routed to and from the control buffers, while data is routed directly into the memory allocated by the task that made the input/output channel request.

At an 800-Mbps datarate, such decisions cannot be made in microcode unless large intermediate buffer areas are used as holding tanks to provide the time needed to parse the header and make a decision. There is very little time to make these decisions, so it is cheaper to make them in hardware than to provide the intermediate buffers and suffer the lag-time penalty of a double transfer from port buffers to user buffers.

The Fibre Channel standards do not describe how ports handle memory management of buffers because this is an internal architecture not visible to other ports. Delay corresponds directly to port congestion, and a congested port is not an efficient one. There are likely to be as many different ways of managing buffers inside a Fibre Channel port as there are in processors.

To aid in the transport of FC-4 data across links, FC-2 defines the following set of building blocks:

1. *Ordered set.* An ordered set consists of four 10-bit characters, a combination of data characters and special characters, that are used to provide certain very low level link functions, such as frame demarcation and signaling between two ends of a link. This signaling provides for initialization of the link after power-on and for certain basic recovery actions.

2. *Frame.* A frame is the smallest indivisible packet of data that is sent on the link. Frame protocol/structure is shown in Figure 8.10. Addressing is done within the frame header. Frames are not visible to the upper-level protocols and consist of the following fields:

   - Start-of-frame delimiter (an ordered set)
   - Frame header (defined by FC-2)

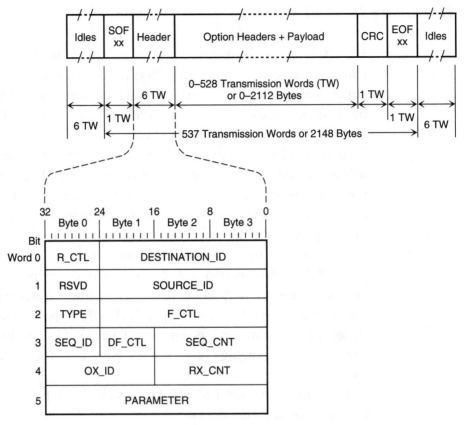

Figure 8.10 *Frame protocol/structure.*

- Optional headers (defined by FC-2)
- Variable-length payload containing FC-4 user data (length 0 to maximum 2112 bytes)
- 32-bit CRC (for error detection)
- End-of-frame delimiter (an ordered set)

Each frame or group of frames is acknowledged as part of the flow-control scheme. The acknowledgments also provide notification of nondelivery of the frame to the N_port at the other end of the link.

Busys and *Rejects* are also defined to provide notification of nondelivery of a frame.

3. *Sequence.* In Fibre Channel, there are no limits on the size of transfers between applications. (In LANs, the software is sensitive to the maximum frame or packet size that can be transmitted.) Frame sizes are transparent to software using Fibre Channel because a buffer, called a *sequence,* is the unit of transfer. A sequence is composed of one or more related frames for a single operation, flowing in the same direction on the link. It is the FC-2 layer's responsibility to break a sequence into the frame size that has been negotiated between the communicating ports and between the ports and the fabric. For example, a unit of data generated by an FC-4 is disassembled into a single sequence of one or more frames. The sequence is also the recovery boundary in Fibre Channel. When an error is detected, Fibre Channel identifies the sequence in error and allows that sequence (and subsequent sequences) to be retransmitted. Each sequence is uniquely identified by the initiator of the sequence via the *Sequence Identifier* (SEQ_ID) field within the frame header. In addition, each frame within the sequence is uniquely numbered with a *Sequence Count* (SEQ_CNT).

4. *Exchange.* An exchange is composed of one or more nonconcurrent sequences for a single operation. For example, an operation may consist of several phases: a command to read some data, followed by the data, followed by the completion status of the operation. Each phase of command, data, and status is a separate sequence, but they can form a single exchange. Within a single exchange, only a single sequence may be active at any one time, although sequences for different exchanges may be concurrently active. This is one form of multiplexing supported by Fibre Channel. The exchange is uniquely identified by each participating N_port.

An *Originator Exchange ID* (OX_ID) is assigned, in an implementation-dependent manner, by the originating (initiator of the first sequence) N_port, and a *Responder Exchange ID* (RX_ID) is assigned, in an implementation-dependent manner, by the responding (recipient of the first sequence) N_port. The exchange IDs are contained within the frame header and are used locally by the

N_ports to manage the exchange however they see fit. An example of exchange usage is shown in Figure 8.11.

Upper Layers

Two other levels of the Fibre Channel architecture are being developed: FC-3 provides common services (for example, striping definition), and FC-4 provides seamless integration of existing standards, accommodating a number of other data communication protocols.

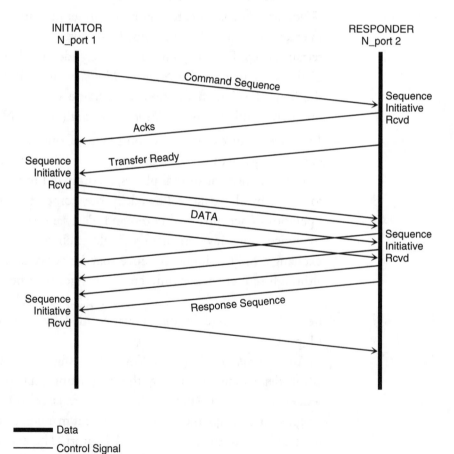

Figure 8.11 Example of exchange usage.

Common Services: FC-3

While the FC-2 level concerns itself with the definition of functions within a single N_port, the FC-3 level concerns itself with functions that span multiple N_ports. To date, the functions defined by FC-3 include the following:

- *Striping.* Striping is a method for achieving higher bandwidth. It uses multiple N_ports in parallel to transmit a single information unit across multiple links and permits multiple links simultaneously.

- *Hunt Groups.* A hunt group is a set of associated N_ports attached to a single node. This set is assigned an alias identifier that allows any frames containing this alias to be routed to any available (that is, nonbusy) N_port within the set. This improves efficiency by decreasing the chance of reaching a busy N_port.

- *Multicast.* Multicast delivers a single transmission to multiple destination N_ports. This includes sending to all N_ports on a fabric (broadcast) or to only a subset of the N_ports on a fabric (multicast).

ULP Mapping: FC-4

FC-4 is the hierarchical level in the Fibre Channel standards suite that specifies the mapping of upper-layer protocols (ULPs) to the levels below, which include FC-3 and FC-PH. Each ULP selected to be transported by Fibre Channel is specified in a separate FC-4 document, as shown in Figure 8.12.

Fibre Channel is equally adept at transporting both network and channel information and allows both protocol types to be concurrently transported over the same physical interface. The following network and channel protocols are currently specified or proposed as FC-4s:

- Small Computer System Interface (SCSI)
- Intelligent Peripheral Interface (IPI)
- High Performance Parallel Interface (HIPPI) Framing Protocol

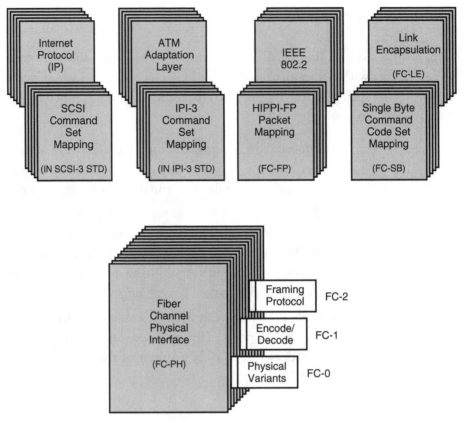

Figure 8.12 *FC-4 protocol mappings.*

- Internet Protocol (IP)
- ATM Adaptation Layer for Computer Data (AAL5)
- Link Encapsulation (FC–LE)
- Single Byte Command Code Set Mapping (SBCCS)
- IEEE 802.2

Proprietary protocols are also possible and permitted but are outside the standard.

The FC–4 mapping protocols make use of the FC–PH physical and signaling protocols to transfer ULP information. Through mapping rules, a specific FC–4 describes how ULP processes of the same FC–4 type interoperate. A networking example of an active ULP process is the transfer of an IS8802-2 PDU from one instance of FC-LE to another. A channel example of a ULP process is a SCSI operation between a channel and a disk drive.

Fibre Channel also provides protocol accelerators for FC-4 usage. Typically, channel protocols follow a command/data/status paradigm. Each of these information categories has different attributes and requires separate processing. However, the processing of each category is common for all protocol types. Common definitions of these information categories allow common implementation across FC-4s.

Fabrics

Fibre Channel employs a switching *fabric* to connect devices, a method that offers numerous advantages. The fabric is similar in many ways to the familiar system used by telephone companies.

For years, telephone companies have provided a complete, low-cost connectivity solution. When a caller picks up a telephone and dials a number, the telephone company routes the call and makes all the intermediate connections needed to ring the number dialed. If the phone is answered, the route is confirmed all the way back to the caller. If a switch fails along the way, the telephone company reroutes the call onto other circuits. Error recovery is not the responsibility of the caller.

In a similar manner, Fibre Channel's fabric relieves each individual port of the responsibility for station management. Each port only has to manage a simple point-to-point connection between itself and the fabric. All the responsibilities for station management associated with routing and error recovery in a network of attachments is, in effect, subcontracted to a switch vendor.

There are no complex routing algorithms to compute to send data to another port far away. The calling port simply does the equivalent of dialing a phone number by entering an identification number for the destination port in the header preceding the actual data. If it is an invalid number, the fabric rejects it. If a congestion problem occurs en route, the fabric responds with a busy signal and the calling port tries again.

The fabric simplifies the traditional means of configuring channels. With a channel, each system vendor has cables routed to devices, and with the limitations on device attachments, a large system winds up with many cables. Distance limitations impose a burden that must be managed by taking extraordinary steps to ensure that the farthest device is within the maximum allowable distance.

The fabric, on the other hand, is a switching mechanism that accepts the responsibility for routing. The only consideration needed for device and processor is the point-to-point attachment of each to the fabric. Distance limits are solved by choosing the media and components needed to meet the application.

The size of a fabric is not physically constrained, but does have an addressing limit of 2^{24} (more than 16 million addresses). A fabric can be as small as a single cable connecting two devices, yet it is also possible to conceive of a heterogeneous mix of Fibre Channel switches (incorporating both circuit and packet switching simultaneously), all cooperating in a very large system in the same way that telephone exchanges operate between countries. Although exchanges are built by different suppliers with a variety of internal architectures, as far as the caller is concerned all telephone calls are made in the same manner. The user is served by an invisible fabric of elements, including local and long-distance components.

There is another reason for adopting principles similar to those practiced by telephone systems. A parallel can be drawn between a LAN and the old telephone party line: The more users on the line, the less bandwidth is available to each. At some point, time available on the line can only be improved by giving each telephone its own line. The Fibre Channel fabric approach does just that. If more bandwidth is needed, more paths are added to the switch, and bandwidth is increased through more parallelism.

As shown in Figure 8.13, the Fibre Channel fabric appears as a single interconnecting entity to the connected N_ports. Internally, a fabric may consist of one or more fabric elements. The primary function of the fabric is to receive frames from a source N_port and to route them to the destination N_port whose address is given in the frame. The fabric provides an address space within the 24-bit address range, and each connected N_port has a unique address identifier. The fabric topology type (defined later), routing path selection within the fabric, and the internal structure of the fabric are transparent to the N_ports. The *F_Port* (fabric port) is the access point of the fabric for physically connecting the user's N_Port. A receiving F_port responds to a sending N_port according to the FC-2 protocol. The F_port may or may not contain a receive buffer for the incoming frames.

Classes of Service

To accommodate a wide range of communications needs, Fibre Channel defines three different classes of service applicable to a fabric and an N_port. A given fabric or N_port may support one or more classes of service.

Class 1, based on a hard- or circuit-switched connection, functions in much the same way as today's dedicated physical channels. An end-to-end path between the communicating devices, through the intervening fabric, is established before any data transfer can begin. When a host and device are linked using Class 1, that path is not available to other hosts. When the time needed to make a connection is short or data transmissions are long, Class 1 is an ideal service. Two interconnected supercomputers would use a Class 1 hookup to ensure rapid and uninterrupted communication.

Class 2 is a frame-switched service without a dedicated connection that provides guaranteed delivery with an acknowledgment of receipt. There is

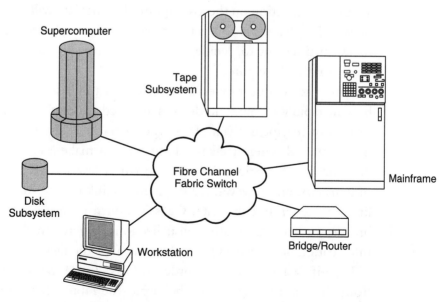

Figure 8.13 *Fibre Channel interconnections through the fabric.*

none of the delay required to establish a connection as in Class 1 and no uncertainty over whether delivery was achieved. If delivery cannot be made due to congestion, a busy signal is returned and the sender tries again. The sender knows that it has to retransmit immediately without waiting for a long time-out to expire. As with traditional packet-switched systems, the path between two interconnected devices is not dedicated, allowing better use of the link's bandwidth. The fabric demultiplexes frames from one or multiple source N_ports to the requested destination N_ports. Class 2 is ideal for data transfers to and from a shared mass storage system physically located at some distance from several individual workstations.

Fibre Channel also provides an optional mode called *Intermix*. Intermix reserves the full Fibre Channel bandwidth for a dedicated (Class 1) connection, but also allows connectionless traffic within the fabric to share the link if the bandwidth is available during idle Class 1 connections. Intermix provides the greatest bandwidth efficiency, enabling a node to maximize its capacity by sharing the unused capacity on a dedicated Class 1 connection.

Distance is the greatest factor influencing connection time. It takes an electrical signal about 10 μs to travel 3 km. To establish a logical connection between two devices separated by this distance requires a signal to make a round trip down and back, about 20 μs, before the transfer can begin. LANs avoid this delay by transferring data without first establishing a logical connection on the assumption that the data will be delivered. The drawback is that if the data is not delivered, the sender does not know, and relies on time-out to decide whether information has been lost or not, retransmitting if required.

Class 3 is a connectionless service, similar to Class 2, that allows data to be sent rapidly to multiple devices attached to the fabric, but no confirmation of receipt is given. This type of transfer, known as a *datagram,* is most practical when it takes a long time to make a connection. By not providing confirmation, the one-to-many Class 3 service speeds the time of transmission. However, if a single user's link is busy, the system will not know to retransmit the data. Class 3 service is very useful for real-time broadcasts, such as weather visualizations, where timeliness is key and information not received has little value after the fact.

Even if data arrives out of order in Class 2 or 3 transmissions, resulting from different routing within the fabric, the Fibre Channel provides controls within the frame to present data in the proper order to the receiving buffer of the application software.

Topologies

Although Fibre Channel shares many features with networks, it differs considerably by the absence of topology dependencies. For example, IEEE 802.6 Token Ring, IEEE 802.2 Ethernet, and ANSI FDDI cannot share the same medium because their station management is topology-dependent.

Fibre Channel is a closed system that relies on ports logging in with each other and with the fabric to trade information on attributes and characteristics so they can decide whether they can work together. If they can work together, they define the criteria under which they will communicate. Whether the fabric is a circuit switch, an active hub, or a loop is irrelevant, because the station management issues related to topology are not part of Fibre Channel ports but, rather, are the responsibility of the fabric.

Selecting the fabric topology is influenced by the system performance requirements, packaging options, and the user's requirement for growth capacity. When combined with choices of physical variants, Fibre Channel offers an unprecedented capability for product differentiation and system tuning to specific applications. However, the key point is that one N_port design is compatible with all fabric implementations. Possible fabric topologies include point-to-point, crosspoint-switched, and arbitrated loop (see Figure 8.14). The remainder of this chapter focuses on crosspoint-switched and arbitrated-loop fabric topologies.

Crosspoint-Switched Topology

The highest-performance Fibre Channel fabric is based on a crosspoint-switched topology (the FC-XS document in development), providing a choice of multiple path routings between pairs of F_ports. This type of operation is similar to the familiar switching telephone exchanges. Multiple moderate-speed paths provide high aggregate system bandwidth. For example, an 8×8 crosspoint switch with 25-Mbps paths, as shown in Figure 8.15, gives aggregate bandwidth of 400 Mbps, with 200 Mbps in each direction using fully duplex communication.

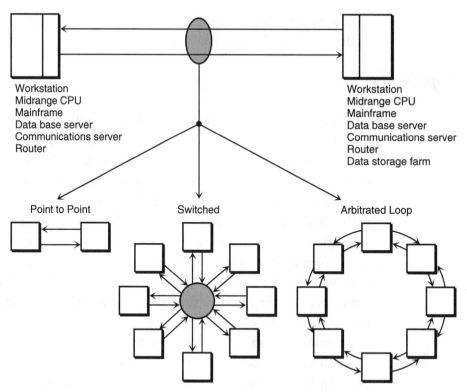

Figure 8.14 *Fibre Channel topologies.*

The crosspoint-switched topology supports all FC-PH-defined classes of service. In the Class 1 circuit-switched mode (private-telephone-line type), the data transfer is based on a dedicated connection, whereas in the frame-switched mode (Classes 2 and 3) the data transfer is connectionless. Figure 8.16 shows the difference between these two modes. In the circuit-switched mode, a dedicated path through the fabric is established on an end-to-end basis between the communicating N_ports before data transfer can begin. The data transfer may be carried out until the path is disconnected by either of the communicating N_ports. Deciding factors for choosing this mode are low circuit set-up time, short message transmission delay, and long transfer.

In the frame switching mode, the bandwidth is dynamically allocated on a link-by-link basis, and the data transmission is in frames. Based on adaptive routing within the fabric, individual frames between the same pair of N_ports are independently switched and may take alternate paths. However, the fabric may be implemented to guarantee in-order delivery

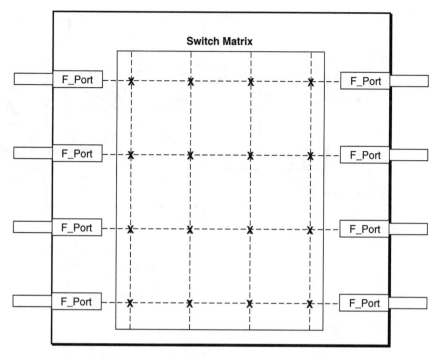

Figure 8.15 *Crosspoint switch.*

of frames to the Destination N_port. During frame switching, buffering is required within the fabric to provide link level flow control between the fabric and the connected N_port. User selection criteria for the frame switching mode are low frame queue delay within the fabric, low frame retransmission probability, and short transfer.

Address Partitioning

The purpose of addressing is to direct the data to the correct location, just as a telephone number directs a call. The address partitioning for the cross-point switch fabric breaks the 24-bit address identifier space into a 3-level hierarchy. The *Domain, Area,* and *Ports* form the highest-, middle-, and lowest-level logical constructs in the addressing hierarchy. Figure 8.17 shows this hierarchy. Again, this is analogous to a telephone call being directed by a telephone number representing a 3-digit area code, 3-digit local exchange code, and 4-digit subscriber number. The address partitioning, which is invisible to the N_ports, aids routing by the fabric.

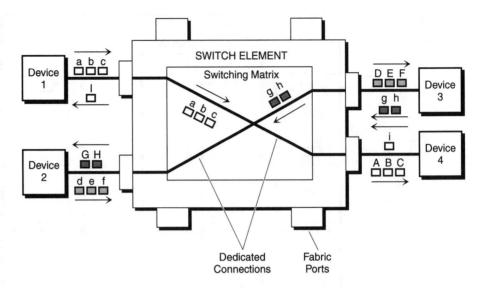

Switching is made for every connection

(a)

Switching is made for every frame—no dedicated connection

(b)

Figure 8.16 Circuit and frame switches.

Fibre Channel fabric provides for automatic address assignment to all the fabric ports as part of an initialization procedure. During the login process between the N_ports and the fabric, the N_ports inherit the addresses of the corresponding fabric ports. Vendor-unique address partitioning and assignment are permitted in a special user environment, but are not part of the FC-XS document.

Routing

Routing within the crosspoint-switched fabric is based on the 24-bit destination address contained within the frame header. Where alternate paths exist between the source and destination ports, the routing path selection is automatic and is based on the lowest- or lower-cost attribute within the fabric. Cost evaluation is based on several factors, such as switch structure, routing ability, bandwidth utilization, service class sharing, flow-control level, and datarate. The switch element periodically reevaluates the path selection, and the reevaluation period is programmable. Vendor-unique routing methods are permitted, but are not part of the FC-XS document.

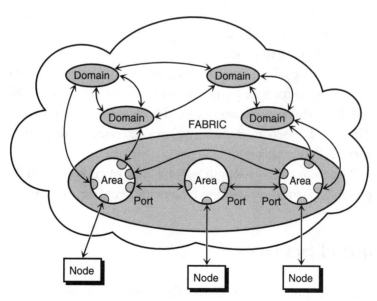

Figure 8.17 *Fabric address partitioning.*

Arbitrated Loop

The Fibre Channel Arbitrated Loop (FC-AL) topology provides a low-cost way to attach multiple communicating ports in a loop without hubs and switches. The loop is a shared-bandwidth, distributed topology where each port includes the minimum necessary connection function. Up to 127 *L_ports* (loop ports) may be in participating mode on the loop. Representative loop configurations are shown in Figure 8.18.

An L_port may arbitrate for use on an arbitrated loop (hereafter referred to as the *loop*). Once an L_port wins the arbitration, based on the lowest port address, a second L_port may be opened to complete a single bidirectional point-to-point circuit. With the loop, only one pair of L_ports may communicate at one time. When the two connected L_ports release control of the loop, another point-to-point circuit may be established. An L_port discovers its environment and works properly (without outside intervention) with an F_port, an N_port, or with another L_port. A loop may be attached to the fabric to connect to ports beyond the loop using an *FL-port*. To connect a number of hosts together without intervening fabric, a host interface called an *NL-port* is used, which combines features of an N_port and an L_port.

The information received on the inbound fiber of the L_port is either directed to the local FC-2 function or placed on the outbound fiber for another L_port to process.

Unlike the fabric topology, where a circuit is established only for a dedicated connection, a point-to-point circuit must be established between two L_ports on the loop before the FC-2 framing protocol may be used. The two L_ports may use any class of service appropriate for a particular user implementation.

The loop is self-configuring and may operate with or without a fabric present. Fairness is provided by guaranteeing equal access of all NL_ports.

Fibre Channel Services

Fibre Channel services, currently in development, provide a set of functions, some of which are required by Fibre Channel protocols and some of

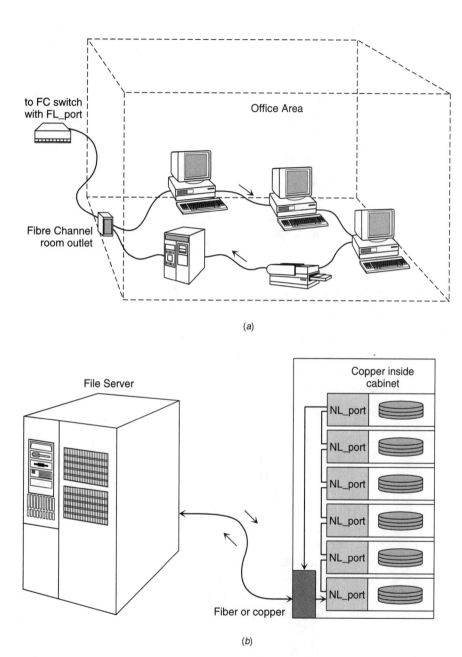

Figure 8.18 Loop configurations: (a) Representative loop, and (b) Disk loop.

which provide optional enhancements to Fibre Channel's basic protocols. These services and their functions are outlined in Table 8.3.

The mandatory services are the following:

- Fabric Controller
- Login Server

The optional services are the following:

- Name Server
- Management Server
- Time Server

Table 8.3 Fibre Channel Server Functions

Login Server
   Helps to discover operating characteristics
   Provides N_port address assignment
   Mandatory within fabric
Fabric Controller
   Assists initialization/configuration
   Routing management
   Optional fabric-assisted services
   Mandatory within fabric
Name Server (optional)
   Translation for N_port IDs
      IEEE/CCITT addresses (worldwide unique)
      IP addresses
      Symbolic names
   Vital product data
Time Server (optional)
   Management of expiration timers
   Time synchronization
Management Server (optional)
   Configuration management
   Performance management
   Fault management
   User diagnostics
   Accounting

Fibre Channel differs from other connection standards in its handling of station management. Other standards require that each individual node handle station-management functions. In Fibre Channel, these functions are handled by special servers within the fabric, and do not have to be handled at each node, resulting in a simpler and less-costly system.

Each server is associated with a well-known, fabric-wide, unique address reserved by Fibre Channel. While the servers appear as N_port addresses, they may also be implemented internal to a fabric. The common transport protocol for all services is the *Fibre Channel Common Transport protocol* (FC-CT). The FC-CT is an N_port-to-N_port protocol that is transparent to the fabric type or topology. The services support all classes of service and both a centralized and a distributed environment.

Fabric Controller

A Fabric Controller provides both required and optional fabric services such as the following:

- Initialization
- Configuration
- Routing management
- Optional fabric-assisted services

Login Server

The Login Server is a logical entity within the fabric that receives and responds to the Fabric Login requests. The Login Server also confirms or assigns the N_port identifier of the N_port that initiates the Login.

Name Server

The primary function of the Name Server is to provide the translation from a given node name to one or more associated N_port identifiers. A node name is an 8-byte value that may be an ITU-T or IEEE address (that is, worldwide unique) or may be locally assigned. In either case, it is unique within fabric address space. N_port identifiers are assigned (optionally by the Login Server) when Login takes place. After this point, a node or N_port may register the association between the node and related N_ports. Translation for objects other than name identifiers and N_port identifiers, such as IP address and upper-level protocols supported, are also provided by the Name Server.

Management Server

The Management Server allows the following types of management:

- Configuration management
- Performance management
- Fault management
- Accounting management
- User diagnostics.

Fibre Channel has adopted the industry-accepted *Simple Network Management Protocol* (SNMP) for providing access to management information. *Management Information Base* (MIB) data are associated with each Fibre Channel node and switch element. SNMP is used to monitor and modify the MIB data. SNMP may use IP and FC-LE over FCPH as its transport mechanism (the conventional use of SNMP). However, not all Fibre Channel nodes will support IP (e.g., peripheral systems). A generic solution is provided in which SNMP uses FC-CT over FC-PH. This allows the management of Fibre Channel networks consisting of both computing equipment and peripheral devices.

Time Server

The Time Server allows for the management of all the timers used within the Fibre Channel system. Other functions of the Time Server are time synchronization of all the switch elements and time stamping of messages generated by the Management Server.

Future Directions: FC-EP and Beyond

Early in the development of Fibre Channel, it became apparent that there was an extremely large set of requirements that needed to be met and that all of the necessary functions could not be initially defined. A subset was therefore chosen on which the Fibre Channel Working Group placed its immediate focus; these are the functions described in FC-PH. Subsequent to the completion of FC-PH, the Fibre Channel Working Group turned its attention to additional physical layer functions that are required in the industry. These functions will become part of the *Fibre Channel Enhanced Physical* (FC-EP) definition, which is the next generation of FC-PH.

The intent of FC-EP is to build on and be compatible with the existing FC-PH. That is, there will be no changes to the existing FC-1 and FC-2 levels. The only changes contemplated to the FC-0 level are the addition of new variants—for example, higher bit rates or support of new modes, such as UTP media and wave-division multiplexing.

In general, two classes of functions are comprehended by FC-EP. The first class includes those functions defined by FC-3 that span multiple N_ports. These include the following:

- *Aliasing.* The FC-EP provides the capability of assigning alias address identifiers to N_ports. An N_port may recognize one or more of these alias IDs, in addition to its native N_port identifier. Alias IDs provide two main functions:

  Hunt groups. As defined, a hunt group allows the fabric to route frames to any member of the hunt group, thereby preventing a busy N_port on a node from blocking communication with the node; the fabric may use another N_port within the hunt group. All the N_Ports within the hunt group are assigned an alias ID representing the hunt group. Any frame containing this alias ID as a destination ID may be delivered by the fabric to any N_port within the hunt group.

  Multicast groups. As defined, multicasting allows a source N_port to send a single frame that is replicated, by the fabric, to all members of the multicast group. All N_ports within a multicast group are assigned an alias ID representing the multicast group. Multicasting is valid only in Class 3; any Class 3 frame containing this alias ID as a destination ID is delivered by the fabric to all N_ports within the multicast group.

In both cases, alias IDs are managed by an alias server, which handles the registration and deregistration of both hunt groups and multicast groups. All routing of the frames for both hunt groups and multicast groups is handled by the fabric. The alias server is involved only in the management of the groups.

The second class includes those functions that extend and enhance the existing FC-0, FC-1, and FC-2 definitions. These include the following:

- *Class 4 service.* Class 4 (Fractional Bandwidth) service is a connection-oriented service for use with a fabric. A Class 4 connection is bidirectional, with one virtual circuit (VC) operational in each direction, and supports a different set of quality of service (QoS) parameters for each VC. These quality of service parameters include guaranteed bandwidth and latency. A quality of service facilitator (QoSF) function is provided within the fabric to manage and maintain the negotiated quality of service on each VC. An N_port may reserve up to 254 concurrent Class 4 connections to the same or different N_ports (see Figure 8.19). Class 4 has been separated into two pieces: the setup of the quality of service parameters on both the VCs to establish the right to the Class 4 connection, and the connection itself. The QoS parameter setup requires a round-trip delay. Thereafter, the resources are available, on call, so participating N_ports can activate or deactivate connections any number of times without paying any turnaround delay penalty. Once a Class 4 connection is active, the fabric paces frames from the source N_port into the fabric on that VC by permitting the source N_port to transmit a frame to the destination N_port. This pacing is the mechanism used by the fabric to regulate available bandwidth per VC. This level of control permits congestion management for a fabric and guarantees access to the destination N_Port. In Class 4, the fabric is responsible for multiplexing frames belonging to different VCs between the same or different N_port pairs. In-order delivery of frames within a VC is guaranteed in Class 4 service.

- *Enhanced Class 1 service.*   This includes two services:

  Dedicated simplex service.   The present Class 1 connection, as defined in FC-PH, is a duplex connection-oriented communication service. FC-EP extends this service to provide simplex unidirectional connection. The dedicated simplex connection service provides a data transmission capability of large block transfers over Class 1 connections uncoupling the inbound and outbound data paths of Class 2 to transfer con-

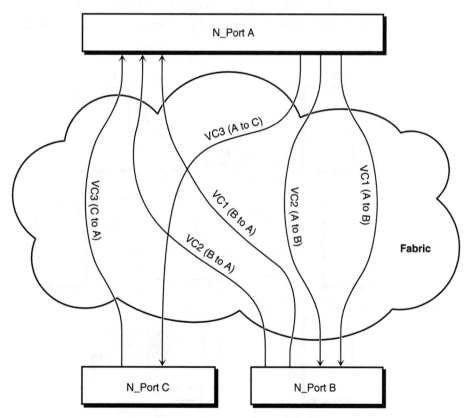

Figure 8.19 *Example of Class 4 connections.*

trol messages such as acknowledgment of receipts (ACKs; see Figure 8.20). The dedicated simplex allows a connection on an outbound link with one N_port while simultaneously having a connection on the inbound link with another N_port. This provides increased bandwidth utilization by allowing the likelihood of simultaneous data transfer over both outbound and inbound links when compared with a Class 1 dedicated connection (duplex).

Buffered Class 1 service.   Buffered Class 1 service is an optional service that allows N_ports of differing data rates to communicate on a Class 1 connection.

- *Data compression.*   FC-EP provides the capability of lossless data compression by accommodating the implementation of the *Adaptive Lossless Data Compression Lempel Ziv-1* (ALDC LZ-1) algorithm as an optional FC-2 function. The option is applicable to all

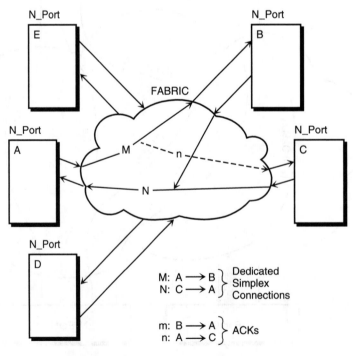

Figure 8.20   Dedicated simplex connections.

classes of service. The option can be exercised by the initiator on a per-information category within a sequence, immediately preceding segmentation to frame payloads. Decompression is performed at the recipient immediately following information category reassembly. This feature can be used in video applications requiring Fibre Channel connection to devices such as digital disk recorders.

- *New FC-0 variants.* New FC-0 variants have been included to provide higher link bit rates in multiples of 1.0625 Gbps.

Industry Information

Fibre Channel will bring to market the high-performance, easy-to-use, low-cost communications pathway required by the new breed of data and communications-intensive applications being developed for business, government, and academic institutions. It will provide high-speed links to massive storage systems at significantly lower costs than today's systems and will enable the construction of heterogeneous clusters of workstations for applications such as medical imaging, visualization, simulation, real-time full-motion video, massive data retrieval, real-time backup, and many others.

The ability of Fibre Channel to replace the variety of separate and distinct interconnects and protocols used by peripherals, coupled with its ability to transport network traffic, allow it to be truly mass-marketed. The resulting large volumes promise to drastically reduce costs, making it possible that Fibre Channel will be the only interface that a system needs for these attachments.

Fibre Channel enables computing systems to transfer data at a rate over 1 Gbps—10 to 250 times faster than today's network protocols—to as many as 16 million nodes. It combines the attributes of a channel with those of a network, a synthesis which results in a lower-cost, simpler, and more efficient solution to many of today's connectivity and internetworking problems.

For more details on current implementations, the Fibre Channel Association can be reached at:

Fibre Channel Association
12407 MoPac Expressway North 100-357
P.O. Box 9700
Austin, TX 78758-9700
Phone: (800) 272-4618
email: FCA-info@amcc.com

For ANSI documentation information, the reader may contact:

Global Engineering
15 Inverness Way East
Englewood, CO 80112-5704
Phone: (800) 854-7179 or (303) 792-2181
Fax: (303) 792-2192

Fibre Channel Glossary

ANSI   American National Standards Institute, the coordinating organization for voluntary standards in the United States.

arbitrated loop topology (FC-AL)   A Fibre Channel topology that provides a low-cost solution to attach multiple communicating ports in a loop without hubs and switches.

arbitration   The process of selecting one respondent from a collection of several candidates that request service concurrently.

ATM   A type of packet switching that transmits fixed-length units of data.

broadcast   Sending a transmission to all N_ports on a fabric.

channel   A point-to-point link, the main task of which is to transport data from one point to another.

controller   A computer module that interprets signals between a host and a peripheral device. The controller often is part of the peripheral device.

CRC   Cyclic redundancy check, an error-correcting code used in Fibre Channel.

crosspoint-switched topology (FC-XS)   Highest-performance Fibre Channel fabric, providing a choice of multiple path routings between pairs of F_ports.

Datagram   The Class 3 Fibre Channel service that allows data to be sent rapidly to multiple devices attached to the fabric, with no confirmation of receipt.

8B/10B   A data encoding scheme developed by IBM, translating byte-wide data to an encoded 10-bit format.

ESCON   Enterprise Systems Connection.

Exchange   One of the Fibre Channel "building blocks," composed of one or more nonconcurrent sequences for a single operation.

Fabric   Fibre Channel–defined interconnection methodology that handles routing in Fibre Channel networks.

FC-EP   The future Fibre Channel Enhanced Physical standard, which will build on and be compatible with FC-PH.

FC-PH   Fibre Channel Physical standard, consisting of the three lower levels, FC-0, FC-1, and FC-2.

FC-0   Lowest level of the Fibre Channel Physical standard, covering the physical characteristics of the interface and media.

FC-1   Middle level of the FC-PH standard, defining the 8B/10B encoding/decoding and transmission protocol.

FC-2   Highest level of FC-PH, defining the rules for the signaling protocol and describing transfer of the frame, sequence, and exchanges.

FC-3   The hierarchical level in the Fibre Channel standard that provides common services, such as striping definition.

FC-4   The hierarchical level in the Fibre Channel standard that specifies the mapping of upper-layer Protocols (ULPs) to levels below.

FDDI   Fiber Distributed Data Interface, ANSI's architecture for a metropolitan area network (MAN); a network based on the use of optical-fiber cable to transmit data at 100 Mbps.

Fibre Channel Service Protocol (FSP)   The common FC-4 level protocol for all services, transparent to the fabric type or topology.

F_port   Fabric port, the access point of the fabric for physically connecting the user's N_port.

frame   A linear set of transmitted bits that define a basic transport element.

HIPPI   High Performance Parallel Interface, an 800-Mbps interface to supercomputer networks (formerly known as high-speed channel), developed by ANSI.

hunt group   A set of associated N_ports attached to a single node, assigned a special identifier that allows any frames containing this identifier to be routed to any available N_port within the set.

Intermix   A mode of service defined by Fibre Channel that reserves the full Fibre Channel bandwidth for a dedicated (Class 1) connection, but also allows connectionless (Class 2) traffic to share the link if the bandwidth is available.

L_port   A Fibre Channel port that supports the arbitrated loop topology.

Link_Control_Facility   A termination card that handles the logical and physical control of the Fibre Channel link for each mode of use.

Login Server   Entity within the Fibre Channel fabric that receives and responds to login requests.

Multicast   Refers to delivering a single transmission to multiple destination N_ports.

N_port   Node port, a Fibre Channel–defined hardware entity at the node end of a link.

Name Server   Provides translation from a given node name to one or more associated N_port identifiers.

network   An aggregation of interconnected nodes, workstations, file servers, and/or peripherals, with its own protocol that supports interaction.

operation   As defined in FC-2, one of the Fibre Channel "building blocks" composed of one or more, possibly concurrent, exchanges.

Ordered Set   A Fibre Channel term referring to four 10-bit characters (a combination of data and special characters) that provide low-level link functions, such as frame demarcation and signaling between two ends of a link. It provides for initialization of the link after power-on and for some basic recovery actions.

Originator   A Fibre Channel term referring to the initiating device.

Port   The hardware entity within a node that performs data communications over the Fibre Channel link.

protocol   A data transmission convention encompassing timing, control, formatting, and data representation.

receiver   A terminal device that includes a detector and signal processing electronics.

Responder   A Fibre Channel term referring to the answering device.

ring   A configuration of computing devices in which all are interconnected in a ring shape; communications between any two points must include all the intermediate points.

Sequence   One of the Fibre Channel "building blocks," made up of one or more related frames for a single operation.

star   In local area networks, a configuration of computing devices in which each user is connected by communication links radiating out of a central hub that handles all communications.

striping   A method for achieving higher bandwidth using multiple N_ports in parallel to transmit a single information unit across multiple levels.

Synchronous Optical Network (SONET)   A standard for optical network elements providing modular building blocks, fixed overheads and integrated operations channels, and flexible payload mappings. Basic level is 51.840 Mbps (OC-1); higher levels are *n* times the basic rate (OC-*n*)

Time Server A Fibre Channel–defined service function that allows for the management of all timers used within a Fibre Channel system.

topology The logical and/or physical arrangement of stations on a network.

transmitter A device that includes a source of driving elements.

Index